Defoe Abbott Marx Hardy Machiavelli Montaigne Chesterton Cooper Emerson Joyce Austen Hugo Eliot Grimm
Melville
Stoker Carroll Christie Maupassant Haggard Byron Schiller Molière
Wilde Garnett Engels Smith Kafka Hall
Goethe Einstein Fitzgerald Hawthorne Dostoyevsky Kipling Doyle Willis
Baum Cotton Henry Nietzsche Balzac
Leslie Dumas Flaubert Turgenev Vatsyayana Crane
Stockton Verne Vinci
Burroughs Gogol Busch
Curtis Tocqueville Whitman
Homer Widger Tolstoy Twain Scott Plato Harte
Darwin Thoreau
Potter Freud Zola Lawrence Dickens Burton Hesse
Kant Jowett Stevenson Cervantes Voltaire
Andersen
London Descartes Wells Cooke
Poe Aristotle Hastings
Hale James Shakespeare Irving
Bunner Chambers
Richter Chekhov da Benedict Alcott
Doré Dante Shaw Pushkin
Swift Wodehouse Newton

tredition

tredition was established in 2006 by Sandra Latusseck and Soenke Schulz. Based in Hamburg, Germany, tredition offers publishing solutions to authors and publishing houses, combined with world-wide distribution of printed and digital book content. tredition is uniquely positioned to enable authors and publishing houses to create books on their own terms and without conventional manufacturing risks.

For more information please visit: www.tredition.com

TREDITION CLASSICS

This book is part of the TREDITION CLASSICS series. The creators of this series are united by passion for literature and driven by the intention of making all public domain books available in printed format again - worldwide. Most TREDITION CLASSICS titles have been out of print and off the bookstore shelves for decades. At tredition we believe that a great book never goes out of style and that its value is eternal. Several mostly non-profit literature projects provide content to tredition. To support their good work, tredition donates a portion of the proceeds from each sold copy. As a reader of a TREDITION CLASSICS book, you support our mission to save many of the amazing works of world literature from oblivion. See all available books at www.tredition.com.

 Project Gutenberg

The content for this book has been graciously provided by Project Gutenberg. Project Gutenberg is a non-profit organization founded by Michael Hart in 1971 at the University of Illinois. The mission of Project Gutenberg is simple: To encourage the creation and distribution of eBooks. Project Gutenberg is the first and largest collection of public domain eBooks.

Trees of the Northern United States Their Study, Description and Determination

A. C. (Austin Craig) Apgar

Imprint

This book is part of TREDITION CLASSICS

Author: A. C. (Austin Craig) Apgar
Cover design: Buchgut, Berlin – Germany

Publisher: tredition GmbH, Hamburg - Germany
ISBN: 978-3-8472-2431-0

www.tredition.com
www.tredition.de

TREES

OF THE

NORTHERN UNITED STATES

THEIR STUDY, DESCRIPTION AND
DETERMINATION
FOR THE
USE OF SCHOOLS AND PRIVATE STUDENTS

BY

AUSTIN C. APGAR

PROFESSOR OF BOTANY IN THE NEW JERSEY STATE NOR-
MAL SCHOOL

"Trees are God's Architecture."—*Anonymous.*

"A Student who has learned to observe and describe so simple a matter as the form of a leaf has gained a power which will be of lifetime value, whatever may be his sphere of professional employment."—*Wm. North Rice.*

NEW YORK-:-CINCINNATI-:-CHICAGO
AMERICAN BOOK COMPANY

W. P. 3. [Pg 3]

PREFACE.

This book has been prepared with the idea that teachers generally would be glad to introduce into their classes work dealing with the real objects of nature, provided the work chosen were of a character that would admit of its being studied at all seasons and in all localities, and that the subject were one of general interest, and one that could be taught successfully by those who have had no regular scientific instruction.

The trees of our forests, lawns, yards, orchards, streets, borders, and parks give us just such a department. Though many consider a large part of the vegetable kingdom of little importance, and unworthy of any serious study, there are few who do not admire, and fewer still who do not desire to know, our trees, the monarchs of all living things.

The difficulty in tree study by the aid of the usual botanies lies mainly in the fact that in using them the first essential parts to be examined are the blossoms and their organs. These remain on the trees a very short time, are often entirely unnoticed on account of their small size or obscure color, and are usually inaccessible even if seen. In this book the leaves, the wood, the bark, and, in an elementary way, the fruit are the parts to which the attention is directed; these all can be found and studied throughout the greater part of the year, and are just the parts that must be thoroughly known by all who wish to learn to recognize trees.

Though every teacher is at liberty to use the book as he thinks best, the author, who has been a class teacher for over twenty years, is of the opinion that but little of Part I. need be [Pg 4] thoroughly studied and recited, with the exception of Chapter III. on leaves. The object of this chapter is not to have the definitions recited (the recitation of definitions in school work is often useless or worse than useless), but to teach the pupil to use the terms properly and to make them a portion of his vocabulary. The figures on pages 38-43 are designed for class description, and for the application of botanical words. The first time the chapter is studied the figure illustrating the term should be pointed out by the pupil; then, as a review of the whole chapter, the student should be required to give a full description of each leaf.

After this work with Chapter III., and the careful reading of the whole of Part I., the pupils can begin the description of trees, and, as the botanical words are needed, search can be made for them under the proper heads or in the Glossary.

The Keys are for the use of those who know nothing of scientific botany. The advanced botanist may think them too artificial and easy; but let him remember that this work was written for the average teacher who has had no strictly scientific training. We can hardly expect that the great majority of people will ever become scientific in any line, but it is possible for nearly every one to become interested in and fully acquainted with the trees of his neighborhood.

The attainment of such botanical knowledge by the plan given in this volume will not only accomplish this useful purpose, but will do what is worth far more to the student, *i.e.*, teach him to employ his own senses in the investigation of natural objects, and to use his own powers of language in their description.

With hardly an exception, the illustrations in the work are taken from original drawings from nature by the author. A few of the scales of pine-cones were copied from London's "Encyclopædia of Trees"; some of the Retinospora cones were taken from the "Gardener's Chronicle"; and three of the illustrations in Part I. are from Professor Gray's works. [Pg 5]

The size of the illustration as compared with the specimen of plant is indicated by a fraction near it; ¼ indicates that the drawing is one fourth as long as the original, 1/1 that it is natural size, etc. The notching of the margin is reduced to the same extent; so a margin which in the engraving looks about entire, might in the leaf be quite distinctly serrate. The only cases in which the scale is not given are in the cross-sections of the leaves among the figures of coniferous plants. These are uniformly three times the natural size, except the section of Araucaria imbricata, which is not increased in scale.

The author has drawn from every available source of information, and in the description of many of the species no attempt whatever has been made to change the excellent wording of such authors as Gray, Loudon, etc.

The ground covered by the book is that of the wild and cultivated trees found east of the Rocky Mountains, and north of the southern boundary of Virginia and Missouri. It contains not only the native species, but all those that are successfully cultivated in the whole region; thus including all the species of Ontario, Quebec, etc., on the north, and many species, both wild and cultivated, of the Southern States and the Pacific coast. In fact, the work will be found to contain so large a proportion of the trees of the Southern States as to make it very useful in the schools of that section.

Many shrubby plants are introduced; some because they occasionally grow quite tree-like, others because they can readily be trimmed into tree-forms, others because they grow very tall, and still others because they are trees in the Southern States.

In nomenclature a conservative course has been adopted. The most extensively used text-book on the subject of Botany, "Gray's Manual," has recently been rewritten. That work includes every species, native and naturalized, of the region covered by this book, and the names as given in that edition have been used in all cases. [Pg 6]

Scientific names are marked so as to indicate the pronunciation. The vowel of the accented syllable is marked by the grave accent (`) if long, and by the acute (´) if short.

In the preparation of this book the author has received much valuable aid. His thanks are especially due to the authorities of the Arnold Arboretum, Boston, Massachusetts, and of the Missouri Botanical Garden, St. Louis, for information in regard to the hardiness of species; to Mr. John H. Redfield, of the Botanical Department of the Philadelphia Academy of Natural Sciences, for books, specimens from which to make illustrations, etc.; and to Dr. A. C. Stokes, of Trenton, New Jersey, for assistance in many ways, but especially for the accurate manner in which he has inked the illustrations from the author's pencil-drawings.

The author also wishes to acknowledge the help received from many nurserymen in gathering specimens for illustration and in giving information of great value. Among these, special thanks are due to Mr. Samuel C. Moon, of Morrisville Nurseries, who placed his large collection of living specimens at the author's disposal, and in many other ways gave him much intelligent aid. [Pg 7]

TREES. [Pg 9]

PART I.

THE ESSENTIAL ORGANS, AND THE TERMS NEEDED FOR THEIR DESCRIPTION.

Chapter I.

Roots.

Though but little study of the roots of trees is practicable, some knowledge of their forms, varieties, and parts is important.

The great office of the roots of all plants is the taking in of food from the soil. Thick or fleshy roots, such as the radish, are stocks of food prepared for the future growth of the plant, or for the production of flowers and fruit. The thick roots of trees are designed mainly for their secure fastening in the soil. The real mouths by which the food is taken in are the minute tips of the hair-like roots found over the surface of the smaller branches. As trees especially need a strong support, they all have either a *tap-root*—one large root extending from the lower end of the trunk deep down into the ground; or *multiple roots*—a number of large roots mainly extending outward from the base of the trunk.

Trees with large tap-roots are very hard to transplant, and cannot with safety be transferred after they have attained any real size. The Hickories and Oaks belong to this class. [Pg 10]

Trees having multiple roots are readily transplanted, even when large. The Maples and Elms are of this class.

Roots that grow from the root-end of the embryo of the seed are called *primary roots*; those growing from slips or from stems anywhere are *secondary roots*.

Some trees grow luxuriantly with only secondary roots; such trees can readily be raised from stems placed in the ground. The Willows and Poplars are good examples of this group. Other trees need all the strength that primary roots can give them; these have to be raised from seed. Peach-trees are specially good examples, but practically most trees are best raised from seed.

A few trees can be easily raised from root-cuttings or from suckers which grow up from roots. The Ailanthus, or "Tree of Heaven," is best raised in this way. Of this tree there are three kinds, two of which have disagreeable odors when in bloom, but the other is nearly odorless. By using the roots or the suckers of the third kind, only those which would be pleasant to have in a neighborhood would be obtained. One of the large cities of the United States has in its streets thousands of the most displeasing of these varieties and but few of the right sort, all because the nurseryman who originally supplied the city used root-cuttings from the disagreeable kind.

If such trees were raised from the seed, only about one third would be desirable, and their character could be determined only when they had reached such a size as to produce fruit, when it would be too late to transplant them. Fruit-trees, when raised from the seed, have to be grafted with the desired variety in order to secure good fruit when they reach the bearing age.

Chapter II. [Pg 11]
Stems and Branches.

The stem is the distinguishing characteristic of trees, separating them from all other groups of plants. Although in the region covered by this book the trees include all the very large plants, size alone does not make a tree.

A plant with a single trunk of woody structure that does not branch for some distance above the ground, is called a *tree*. Woody plants that branch directly above the soil, even though they grow to the height of twenty feet or more, are called *shrubs*, or, in popular language, *bushes*. Many plants which have a tendency to grow into the form of shrubs may, by pruning, be forced to grow tree-like; some that are shrubs in the northern States are trees further south.

All the trees that grow wild, or can be cultivated out of doors, in the northern States belong to one class, the stems having a separable bark on the outside, a minute stem of pith in the center, and, between these, wood in annual layers. Such a stem is called *exogenous* (outside-growing), because a new layer forms on the outside of the wood each year.

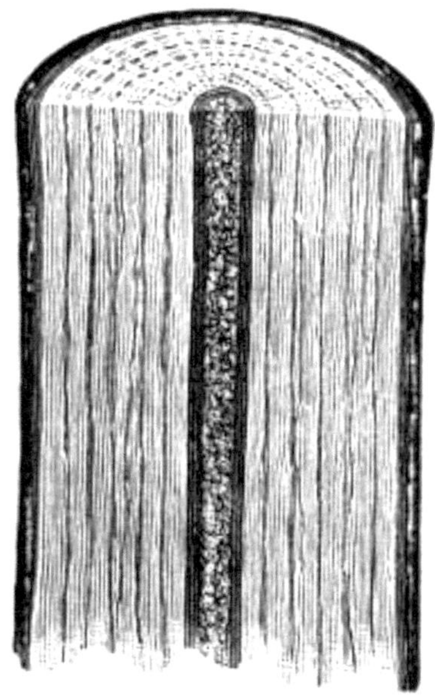

Fig. 1.

Another kind of tree-stem is found abundantly in the tropics; one, the Palmetto, grows from South Carolina to Florida. While in our region there are no trees of this character, there are plants having this kind of stem, the best illustration being the corn-stalk. In this case there is no separable bark, and the woody substance is in threads within the pithy material. In the corn-stalk the woody threads are not very numerous, and the pith is very abundant; in most of the tropical trees belonging to this group the threads of wood are so numerous as to make the ma [Pg 12] terial very durable and fit for furniture. A stem of this kind is called *endogenous* (inside-growing). Fig. 1 represents a longitudinal and a cross section of an exogenous stem, and Fig. 2 of an endogenous one.

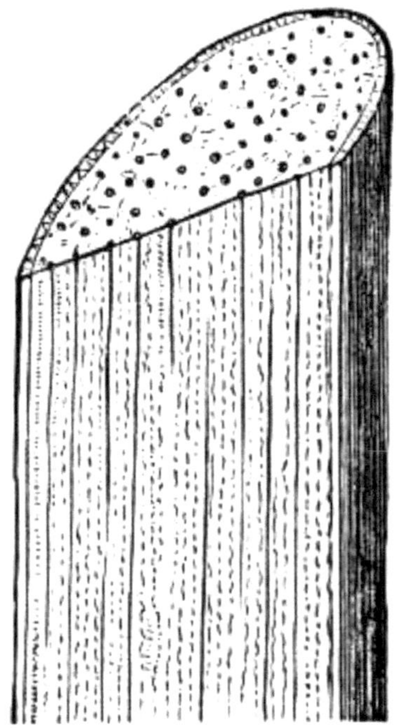

Fig. 2.

Since all the stems with which we have to deal are exogens, a particular description of that class will here be given. Fig. 1 shows the appearance of a section of an Ash stem six years old. The central portion, which is about as thick as wrapping-twine, is the *pith*; from this outward toward the bark can be seen the six annual layers of the *wood*; and then comes the *bark*, consisting of two portions. First there is an inside layer of greenish material, the fresh-growing portion, and lastly the outer or dead matter. This outer portion must crack open, peel off, or in some way give a chance for the constant growth of the trunk. The different kinds of trees are readily known by the appearance of the bark of the trunk, due to the many varieties of surface caused by the allowance for growth. None of the characteristics of trees afford a better opportunity for careful observation and study than the outer bark.

The Birches have bark that peels off in thin horizontal layers—the color, thinness, and toughness differing in the different species; the Ashes have bark which opens in many irregular, netted cracks moderately near each other; the bark of the Chestnut opens in large longitudinal cracks quite distant from one another. The color of the bark and the character of the scales are quite different in the White and the Black Oaks.

In the woody portion radiating lines may be seen; [Pg 13] these are the *silver grain*; they are called by the botanist *medullary rays*.

The central portion of the wood of many large stems is darker in color than the rest. This darker portion is dead wood, and is called *heart-wood*; the outer portion, called *sap-wood*, is used in carrying the sap during the growing season. The heart-wood of the Walnut-tree is very dark brown; that of the Cherry, light red; and that of the Holly, white and ivory-like. The heart-wood is the valuable part for lumber.

If examined under a magnifying glass, the *annual layers* will be seen to consist of minute tubes or cells. In most trees these tubes are much larger in the portion that grew early in the season, while the wood seems almost solid near the close of the annual layer; this is especially true in the Ashes and the Chestnut; some trees, however, show but little change in the size of the cells, the Beech being a good example. In a cross-section, the age of such trees as the Chestnut can readily be estimated, while in the Beech it is quite difficult to do this. Boxwood, changing least in the character of its structure, is the one always used for first-grade wood-engravings.

When wood is cut in the direction of the silver grain, or cut "quartering" as it is called by the lumbermen, the surface shows this cellular material spread out in strange blotches characteristic of the different kinds of wood. Fig. 16 shows an Oak where the blotches of medullary rays are large. In the Beech the blotches are smaller; in the Elm quite small. Lumber cut carefully in this way is said to be "quartered," and with most species its beauty is thereby much increased.

Any one who studies the matter carefully can become acquainted with all the useful and ornamental woods used in a region; the differences in the color of the heart-wood, the character of the annual layers, and the size and the distribution of the medullary rays, afford enough peculiarities to distinguish any one from all others. [Pg 14]

Branching. — The regular place from which a branch grows is the *axil* of a leaf, from what is called an *axillary bud*; but branches cannot grow in the axils of all leaves. A tree with opposite leaves occasionally has opposite branches; while a tree with alternate leaves has all its branches alternate.

Most branches continue their growth year after year by the development of a bud at the end, called a *terminal bud*. Many trees form this bud for the next year's growth so early in the year that it is seldom or never killed by the winter weather; such trees grow very regularly and are symmetrical in form. Most evergreens are good examples. Fig. 3 represents a good specimen. The age of such trees, if not too great, can be readily ascertained by the regularity of each year's growth. The tree represented is sixteen years old. The branches that started the fifth year, about the age at which regular growth begins, are shown by their scars on the trunk.

Fig. 3.

The terminal buds of many trees are frequently killed by the frosts of winter; such trees continue their growth by the development of axillary buds; but as growth from an axillary bud instead of a terminal one will make a branch crooked, such trees are irregular in their branching and outline. Just which axillary buds are most apt to grow depends upon the kind of tree, but trees of the same variety are nearly uniform in this respect. Most trees are therefore readily recognized by the form of outline and the characteristic branching. A good example of a tree [Pg 15] of very irregular growth is the Catalpa (Indian Bean), shown in Fig. 4. The tendency to grow irregularly usually increases with age. The Buttonwood, for example, grows quite regularly until it reaches the age of thirty to forty years; then its new branches grow in peculiarly irregular ways. The twigs of a very old and a young Apple-tree illustrate this change which age produces.

Fig. 4.

There are great differences in the color and surface of the bark of the twigs of different species of trees; some are green (Sassafras), some red (Peach, on the sunny side), some purple (Cherry). Some are smooth and dotless, some marked with dots (Birch), some roughened with corky ridges (Sweet Gum), etc.

The taste and odor of the bark are characteristics worthy of notice: the strong, fragrant odor of the Spice-bush; the fetid odor of the Papaw; the aromatic taste of the Sweet Birch; the bitter taste of the Peach; the mucilaginous Slippery Elm; the strong-scented, resinous, aromatic Walnut, etc.

The branches of trees vary greatly in the thickness of their tips and in their tendency to grow erect, horizontal, or drooping. Thus the delicate spray of the Birches contrasted with the stout twigs of the Ailanthus, or the drooping twigs of the Weeping Willow with the erect growth of the Lombardy Poplar, give contrasts of the strongest character. In the same way, the directions the main branches take in their growth from the [Pg 16] trunk form another distinctive feature. Thus the upward sloping branches of the Elm form a striking contrast to the horizontal or downward sloping branches of the Sour Gum, or, better still, to certain varieties of Oaks.

When the main trunk of a tree extends upward through the head to the tip, as in Fig. 3, it is said to be *excurrent*. When it is soon lost in the division, as in Fig. 4, it is said to be *deliquescent*.

Chapter III. [Pg 17]
Leaves.

Leaves are the lungs of plants. The food taken in by the roots has to pass through the stem to the leaves to be acted upon by the air, before it becomes sap and is fit to be used for the growth of the plant. No portion of a plant is more varied in parts, forms, surface, and duration than the leaf.

No one can become familiar with leaves, and appreciate their beauty and variety, who does not study them upon the plants themselves. This chapter therefore will be devoted mainly to the words needed for leaf description, together with their application.

The Leaf.—In the axil of the whole leaf the bud forms for the growth of a new branch. So by noting the position of the buds, all the parts included in a single leaf can be determined. As a general thing the leaf has but one blade, as in the Chestnut, Apple, Elm, etc.; yet the Horse-chestnut has 7 blades, the Common Locust often has

21, and a single leaf of the Honey-locust occasionally has as many as 300. Figs. 17-58 (Chapter VII.) are all illustrations of single leaves, except Fig. 43, where there are two leaves on a twig. A number of them show the bud by which the fact is determined (Figs. 25, 26, 31, 33, 34, 36, 40, etc.); others show branches which grew from the axillary buds, many of them fruiting branches (Figs. 37, 42, 43, 50, and 54), one (Fig. 51) a thorny branch.

The cone-bearing plants (Figs. 59-67) have only simple leaves. Each piece, no matter how small and scale-like, may have a branch growing from its axil, and so may form a whole leaf. A study of these figures, together [Pg 18] with the observation of trees, will soon teach the student what constitutes a leaf.

Fig. 5.

Arrangement.—There are several different ways in which leaves are arranged on trees; the most common plan is the *alternate* ; in this only one leaf occurs at a joint or node on the stem. The next in frequency is the *opposite* , where two leaves opposite each other are found at the node. A very rare arrangement among trees, though common in other plants, is the *whorled* , where more than two leaves, regularly arranged around the stem, are found at the node. When a number of leaves are bundled together,—a plan not rare among evergreens,—they are said to be *fasciculated* or in *fascicles* .

The term *scattered* is used where alternate leaves are crowded on the stem. This plan is also common among evergreens.

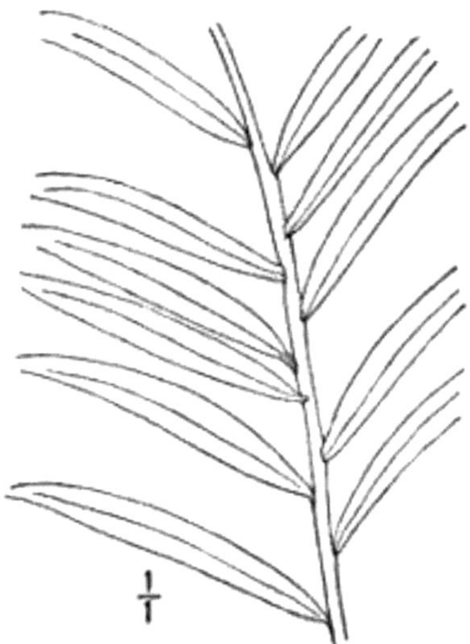

Fig. 6.

Caution. — In some plants the leaves on the side shoots or spurs of a twig are so close together, the internodes being so short, that at first sight they seem opposite. In such cases, the leaf-scars of the preceding years, or the arrangement of the branches, is a better test of the true arrangement of the leaves. The twig of Birch shown in Fig. 5 has alternate leaves.

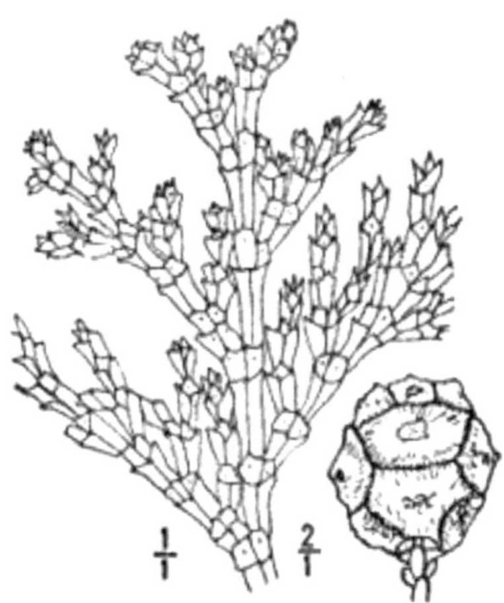

Fig. 7.

There is one variety of alternation, called *two-ranked*, which is quite characteristic of certain trees; that is, the leaves are so flattened out as to be in one plane on the opposite sides of the twig (Fig. 6). The Elm-trees form good examples [Pg 19] of two-ranked alternate leaves, while the Apple leaves are alternate without being two-ranked. Most leaves spread from the stem, but some are *appressed*, as in the Arbor-vitæ (Fig. 7). In this species the *branches* are *two-ranked*.

Parts of Leaves.—A *complete leaf* consists of three parts: the *blade*, the thin expanded portion; the *petiole*, the leafstalk; and the *stipules*, a pair of small blades at the base of the petiole. The petiole is often very short and sometimes wanting. The stipules are often absent, and, even when present, they frequently fall off as soon as the leaves expand; sometimes they are conspicuous. Most Willows show the stipules on the young luxuriant growths.

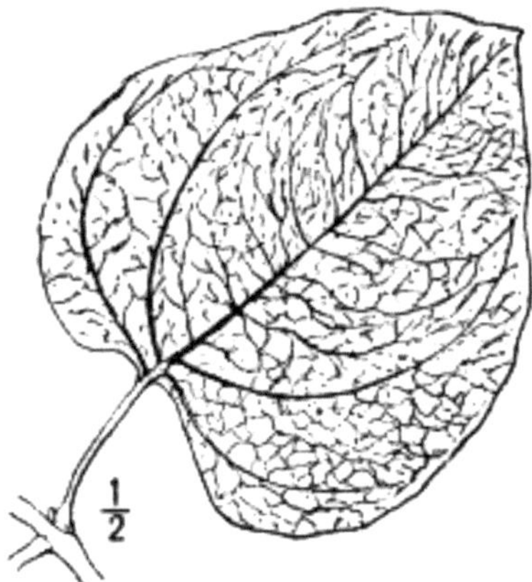

Fig. 8.

Veining.—The leaves of most trees have a distinct framework, the central line of which is called a *midrib*; sometimes the leaf has several other lines about as thick as the midrib, which are called *ribs*; the lines next in size, including all that are especially distinct, are called *veins*, the most minute ones being called *veinlets* (Fig. 8).

Kinds.—Leaves are *simple* when they have but one blade; *compound* when they have more than one. Compound leaves are *palmate* when all the blades come from one point, as in the Horse-chestnut; *pinnate* when they are arranged along the sides, as in the Hickory. Pinnate leaves are of two kinds: *odd-pinnate* , [Pg 20] when there is an odd leaflet at the end, as in the Ash, and *abruptly pinnate* when there is no end leaflet.

Many trees have the leaves *twice pinnate*; they are either *twice odd-pinnate* or *twice abruptly pinnate* . The separate blades of a compound leaf are called *leaflets*. Leaves or leaflets are *sessile* when they have no stems, and *petiolate* when they have stems.

When there are several ribs starting together from the base of a blade, it is said to be *radiate-veined* or *palmate-veined* . When the great veins all branch from the midrib, the leaf is *feather-veined* or *pinnate-veined* . If these veins are straight, distinct, and regularly placed, the leaf is said to be *straight-veined* . The Chestnut is a good example. Leaves having veinlets joining each other like a net are said to be *netted-veined*. All the trees with broad leaves in the northern United

States, with one exception, have netted-veined foliage. A leaf having its veinlets parallel to one another is said to be *parallel-veined* or -*nerved* . The Ginkgo-tree, the Indian Corn, and the Calla Lily have parallel-veined leaves. The narrow leaves of the cone-bearing trees are also parallel-veined.

Forms.—Leaves can readily be divided into the three following groups with regard to their general outline:

1. *Broadest at the middle. Orbicular* , about as broad as long and rounded. *Oval* , about twice as long as wide, and regularly curved. *Elliptical* , more than twice as long as wide, and evenly curved. *Oblong* , two or three times as long as wide, with the sides parallel. *Linear* , elongated oblong, more than three times as long as wide. *Acerose* , needle-shaped, like the leaf of the Pine-tree. [Pg 21]

2. *Broadest near the base. Deltoid* , broad and triangular. *Ovate* , evenly curved, with a broad, rounded base. *Heart-shaped* or *cordate* , similar to ovate, but with a notch at the base. *Lanceolate* , shaped like the head of a lance. *Awl-shaped* , shaped like the shoemaker's curved awl. *Scale-shaped* , short, rounded, and appressed to the stem. The Arbor-vitæ has both awl-shaped and scale-shaped leaves.

3. *Broadest near the apex. Obovate* , same as ovate, but with the stem at the narrow end. *Obcordate* , a reversed heart-shape. *Oblanceolate* , a reversed lanceolate. *Wedge-shaped* or *cuneate* , having a somewhat square end and straight sides like a wedge.

These words are often united to form compound ones when the form of the leaf is somewhat intermediate. The term which most nearly suits the general form is placed at the end; thus *lance-ovate* indicates a leaf between lanceolate and ovate, but nearer ovate than lanceolate; while *ovate-lanceolate* indicates one nearer lanceolate.

Bases.—Oftentimes leaves are of some general form, but have a peculiar base, one that would not be expected from the statement of shape. An ovate leaf which should have a rounded base might have a tapering one; it would then be described as ovate with a *tapering base* . A lanceolate leaf should naturally have a tapering base, but might have an *abrupt* one. Many leaves, no matter what their general form may be, have more or less notched bases; such bases are called *cordate* , *deeply* or *slightly*, as the case may be; and if the lobes at base are elongated, *auriculate* . If the basal lobes project outward, the term *halberd-shaped* is used. Any form of leaf may have a base more or less *oblique* . [Pg 22]

Points. — The points as well as the bases of leaves are often peculiar, and need to be described by appropriate terms. *Truncate* indicates an end that is square; *retuse* , one with a slight notch; *emarginate*, one with a decided notch; *obcordate*, with a still deeper notch; *obtuse* , angular but abrupt; *acute* , somewhat sharpened; *acuminate* , decidedly sharp-pointed; *bristle-pointed* and *awned* , with a bristle-like tip; *spiny-pointed*, with the point sharp and stiff (Holly); *mucronate* , with a short, abrupt point.

Margins. — *Entire* , edge without notches; *repand* , slightly wavy; *sinuate* , decidedly wavy; *dentate* , with tooth-like notches; *serrate* , with notches like those of a saw; *crenate* , with the teeth rounded; *twice serrate* , when there are coarse serrations finely serrated, as on most Birch leaves; *serrulate*, with minute serrations; *crenulate*, with minute crenations. Leaves can be *twice crenate* or *sinuate-crenate*. *Revolute* indicates that the edges are rolled over.

When a leaf has a few great teeth, the projecting parts are called *lobes*, and the general form of the leaf is what it would be with the notches filled in. In the description of such leaves, certain terms are needed in describing the plan of the notches, and their depth and form.

Leaves with palmate veining are *palmately lobed* or *notched*; those with pinnate veining are *pinnately lobed* or *notched*. While the term *lobe* is applied to all great teeth of a leaf, whether rounded or pointed, long or short, still there are four terms sometimes used having special signification with reference to the depth of the notches. *Lobed* indicates that the notches extend about one fourth the distance to the base or midrib; *cleft*, that they extend one half the [Pg 23] way; *parted*, about three fourths of the way; and *divided*, that the notches are nearly deep enough to make a compound leaf of separate leaflets.

So leaves may be palmately lobed, cleft, parted or divided, and pinnately lobed, cleft, parted or divided. The term *pinnatifid* is often applied to pinnately cleft leaves. The terms *entire, serrate, crenate, acute-pointed*, etc., are applied to the lobes as well as to the general margins of leaves.

Surface. — The following terms are needed in describing the surface of leaves and fruit.

Glabrous, smooth; *glaucous*, covered with a whitish bloom which can be rubbed off (Plum); *rugous*, wrinkled; *canescent*, so covered with minute hairs as to appear silvery; *pubescent*, covered with fine, soft, plainly seen hairs; *tomentose*, densely covered with matted hairs; *hairy*, having longer hairs; *scabrous*, covered with stiff, scratching points; *spiny*, having stiff, sharp spines; *glandular-hairy*, having the hairs ending in glands (usually needing a magnifying glass to be seen).

Texture. — *Succulent*, fleshy; *scarious*, dry and chaffy; *punctate*, having translucent glands, so that the leaf appears, when held toward the light, as though full of holes; *membranous*, thin, soft, and rather translucent; *thick, thin*, etc.

Duration. — *Evergreen*, hanging on the tree from year to year. By noticing the color of the different leaves and their position on the twigs, all evergreen foliage can readily be determined at any time during the year. *Deciduous*, falling off at the end of the season. *Fugacious*, falling early, as the stipules of many leaves.

Chapter IV. [Pg 24]
Flowers and Fruit.

The author hopes that those who use this work in studying trees will become so much interested in the subject of Botany as to desire more information concerning the growth and reproduction of plants than can here be given. In Professor Asa Gray's numerous works the additional information desired may be obtained: "How Plants Grow" contains an outline for the use of beginners; "The Elements of Botany" is a more advanced work; while the "Botanical Text Book", in several volumes, will enable the student to pursue the subject as far as he may wish. In this small book the barest outline of the parts of flowers and fruit and of their uses can be given.

Fig. 9.

Flowers.—Parts. The flowers of the Cherry or Apple will show the four kinds of organs that belong to a complete flower. Fig. 9 represents an Apple-blossom. The *calyx* is the outer row of leaves, more or less united into one piece. The *corolla* is the row of leaves within the calyx; it is usually the brightest and most conspicuous part of the flower. The *stamens* are the next organs; they are usually, as in this case, small two-lobed bodies on slender, thread-like stalks. The enlarged parts contain a dust-like material called *pollen*. [Pg 25] The last of the four kinds of parts is found in the center of the flower, and is called the *pistil*. It is this part which forms the fruit and incloses the seed.

The stamens and the pistil are the *essential* organs of a flower, because they, and they only, are needed in the formation of seeds. The pollen from the stamen, acting on the pistil, causes the *ovules* which are in the pistil to grow into *seeds*.

The calyx and corolla are called *enveloping organs*, since they surround and protect the essential parts.

The pieces of which the calyx is composed are called *sepals*. The Apple-blossom has five sepals.

The pieces that compose the corolla are called *petals*.

Fig. 10.

Kinds of Flowers.—When the petals are entirely separate from each other, as in the Apple-blossom, the flower is said to be *polypetalous*; when they grow together more or less, as in the Catalpa (Fig. 10), *monopetalous*; and when the corolla is wanting, as in the flowers of the Oak, *apetalous*.

When all sides of a flower are alike, as in the Apple-blossom, the flower is *regular*; when one side of the corolla differs from the other in color, form, or size, as in the Common Locust, or Catalpa, the flower is *irregular*.

In trees the stamens and pistils are often found in separate flowers; in that case the blossoms containing stamens are called *staminate*, and those containing pistils *pistillate*; those that contain both are called *perfect*. Staminate and pistillate flowers are usually found on the same tree, as in the Oaks, Birches, Chestnut, etc.; in that case the plant is said to be *monœcious*, and all trees of this kind produce fruit. Sometimes, however, the staminate and pistillate flowers are on separate trees, as in the Willows, which are *diœcious*; and then only a portion of the trees—those with pistillate flowers—produce fruit. [Pg 26]

Arrangement of Flowers.—Flowers, either solitary or clustered, grow in one of two ways; either at the end of the branches, being then called *terminal*, or in the axils of the leaves, then called *axillary*. The stem of a solitary flower or the main stem of a cluster is called a *peduncle*; the stems of the separate blossoms of a cluster are called *pedicels*. When either the flowers or the clusters are without stems, they are said to be *sessile*.

Clusters with Pedicellate Flowers.

Raceme , flowers on pedicels of about equal length, scattered along the entire stem. Locust-tree.

Corymb , like a raceme except that the lower flowers have longer stems, making the cluster somewhat flat-topped; the outer flowers bloom first. Hawthorn.

Cyme , in appearance much like a corymb, but it differs in the fact that the central flower blooms first. Alternate-leaved Cornel.

Umbel , stems of the separate flowers about equal in length, and starting from the same point. Garden-cherry.

Panicle , a compound raceme. Catalpa.

Thyrsus, a compact panicle. Horse-chestnut.

Clusters with Sessile or Nearly Sessile Flowers.

Catkin , bracted flowers situated along a slender and usually drooping stem. This variety of cluster is very common on trees. The Willows, Birches, Chestnuts, Oaks, Pines, and many others have their flowers in catkins.

Head , the flowers in a close, usually rounded cluster. Flowering Dogwood.

Fruit. — In this book a single fruit will include all the parts that grow together and contain seeds, whether from [Pg 27] a single blossom or a cluster; there will be no rigorous adherence to an exact classification; no attempt made to distinguish between fruits formed from a simple pistil and those from a compound one; nor generally between those formed from a single and those formed from a cluster of flowers. The fruit and its general classification, determined by the parts easily seen, is all that will be attempted.

As stated before, it is hoped that this volume will not end the student's work in the investigation of natural objects, but that the amount of information here given will lead to the desire for much more.

Berry will be the term applied to all fleshy fruits with more than one seed buried in the mass. Persimmon, Mulberry, Holly. The *pome* or *Apple-pome* differs from the berry in the fact that the seeds are situated in cells formed of hardened material. Apple, Mountain-ash. The *Plum* or *Cherry drupe* includes all fleshy fruits with a single stony-coated part, even if it contains more than one seed. Peach, Viburnum, China-tree. In some cases, when there is but one seed in the flesh and that not stony-coated, it will be called a *drupe-like berry*.

The *dry drupe* is like the Cherry drupe except that the flesh is much harder. The fruit of the Walnut, Hickory, and Sumac.

Fig. 11.

The inner hard-coated parts of these and some others will be called *nuts*. If the nut has a partial scaly covering, as in the Oaks, the whole forms an *acorn* . If the coating has spiny hairs, as in the Chestnut and Beechnut, the whole is a *bur*. The coating in these cases is an *involucre*. If the coating or any part of the fruit has a regular place for splitting open, it is [Pg 28] *dehiscent* (Chestnut, Hickory-nut); if not, *indehiscent* (Black Walnut).

Fig. 12.

Dry fruits with spreading, wing-like appendages, as in the Ash (Fig. 11), Maple (Fig. 12), Elm (Fig. 13), and Ailanthus, are called *samaras* or *keys*.

Dry fruits, usually elongated, containing generally several seeds, are called *pods*. If there is but one cell and the seeds are fastened along one side, *Pea-like pods*, or *legumes*. Locust. The term *capsule* indicates that there is more than one cell. Catalpa, Hibiscus.

Fig. 13.

All the dry, scaly fruits, usually formed by the ripening of some sort of catkin of flowers, will be included under the term *cone* . Pine, Alder, Magnolia. If the appearance of the fruit is not much different from that of the cluster of flowers, as in the Hornbeams, Willows, and Birches, the term *catkin* will be retained for the fruit also. The scales of a cone may lap over each other; they are then said to be *imbricated* or *overlapping* , (Pine); or they may merely touch at their edges, when they are *valvate* (Cypress). When cones or catkins hang downward, they are *pendent*. If the scales have projecting points, these points are *spines* if strong, and *prickles* if weak. The parts back of the scales are *bracts*; these often project beyond the scales, when they are said to be *exserted* . Sometimes the exserted bracts are bent backward; they are then said to be *recurved* or *reflexed.*

Chapter V. [Pg 29]
Winter Study of Trees.

Many of the peculiarities of trees can be studied much better during the winter and early spring than at any other time of the year. The plan of branching, the position, number, size, form, color, and surface of buds, as well as the arrangement of the leaves within the bud and the peculiarities of the scales that cover them, are points for winter investigation.

General Plan of Branching.—There are two distinct and readily recognized systems of branching. 1. The main stem is *excurrent* (Fig. 3) when the trunk extends as an undivided stem throughout the tree to the tip; this causes the spire-like or conical trees so common among narrow-leaved evergreens. 2. The main stem is *deliquescent* (Fig. 4) when the trunk divides into many, more or less equal divisions, forming the broad-topped, spreading trees. This plan is the usual one among deciduous trees. A few species, however, such as the Sweet Gum and the Sugar-maple, show the excurrent stem while young, yet even these have a deliquescent stem later in life. The English Maple and the Apple both have a deliquescent stem very early.

All the narrow-leaved evergreens, and many of the broad-leaved trees as well, show what is called *definite* annual growths; that is, a certain amount of leaf and stem, packed up in the winter bud, spreads out and hardens with woody tissue early in the year, and then, no matter how long the season remains warm, no additional leaves or stem will grow. The buds for the next year's [Pg 30] growth then form and often become quite large before autumn.

There are many examples among the smaller plants, but rarely one among the trees, of *indefinite* annual growth; that is, the plant puts forth leaves and forms stems throughout the whole growing-season. The common Locust, the Honey-locust, and the Sumacs are illustrations.

Buds.—Buds are either undeveloped branches or undeveloped flowers. They contain within the scales, which usually cover them, closely packed leaves; these leaves are folded and wrinkled in a number of different ways that will be defined at the end of this chapter.

Fig. 14.

Position and Number.—While the axils of the leaves and the ends of the stems are the ordinary places for the buds, there are many peculiarities in regard to their exact position, number, etc., that render them very interesting for winter study. Sometimes there are several to the single leaf. In the Silver Maple there are buds on each side of the true axillary one; these are flower-buds, and during the winter they are larger than the one which produces the branch. The Butternut (Fig. 14) and the Walnut have several above each other, the upper one being the largest and at quite a distance from the true axil. In these cases the uppermost is apt to grow, and then the branch is said to be *extra-axillary*. In the Sycamore the bud does not show while the leaf remains on the tree, as it is in the hollow of the leafstalk. In the winter the bud has a ring-like scar entirely around it, instead of the moon-shaped scar below as in most trees. The Common Locust has several buds under the leafstalk and one above it in [Pg 31] the axil. This axillary bud may grow during the time the leaf remains on the tree, and afterward the growth of the strongest one of the others may give the tree two branches almost together.

Some plants form extra buds especially when they are bruised or injured; those which have the greatest tendency to do so are the Willows, Poplars, and Elms. Such buds and growths are called *adventitious*. By cutting off the tops or *pollarding* such trees, a very great number of adventitious branches can be made to grow. In this way the Willow-twigs used for baskets are formed. Adventitious buds form the clusters of curious thorns on the Honey-locust and the tufts of whip-like branches on the trunks and large limbs of the Elms.

In trees the terminal bud and certain axillary ones, differing according to the species or variety of tree, are, during the winter, much larger than the rest. These are the ones which naturally form the new growth, and upon their arrangement the character of branching and thus the form of the tree depend. Each species has some peculiarity in this regard, and thus there are differences in the branching of all trees. In opposite-leaved plants the terminal bud may be small and weak, while the two buds at its side may be strong and apt to grow. This causes a forking of the branches each year. This plan is not rare among shrubs, the Lilac being a good example.

Bud-Scales.—The coverings of buds are exceedingly varied, and are well worthy of study and investigation. The large terminal buds of the Horse-chestnut, with their numerous scales, gummy on the outside to keep out the dampness, and hairy within to protect them from sudden changes of temperature, represent one extreme of a long line; while the small, naked, and partly buried buds of the Honey-locust or the Sumac represent the other end.

The scales of many buds are merely extra parts formed for their protection, and fall immediately after the burst [Pg 32] ing of the buds; while other buds have the stipules of the leaves as bud-scales; these remain on the twigs for a time in the Tulip-tree, and drop immediately in the Magnolia.

Forms of Buds.—The size of buds varies greatly, as before stated, but this difference in size is no more marked than the difference in form. There is no better way to recognize a Beech at any time of the year than by its very long, slender, and sharp-pointed buds. The obovate and almost stalked buds of the Alders are also very conspicuous and peculiar. In the Balsam Poplar the buds are large, sharp-pointed, and gummy; in the Ailanthus they cannot be seen.

Fig. 15.

All the things that might be learned from a small winter twig cannot be shown in an engraving, but the figures here given illustrate some of the facts easily determined from such specimens. The first twig (Ash) had opposite leaves and is 3 years old (the end of each year's growth is marked by dotted lines on all the figures); the year before last it had 6 leaves on the middle portion; last year it had 8 leaves on the end portion and 12 on the side shoots of the middle portion. The buds near the [Pg 33] end of the annual growth are strongest and are most apt to grow. The specimen illustrated was probably taken from the end of a branch of a rather young and luxuriantly growing tree. Thus the Ash must have quite a regular growth and form a regularly outlined tree.

The second twig (Sweet Gum) shows 7 years' growth and is probably a side shoot from more or less within the tree-top. It is stunted in its growth by the want of light and room. The leaves were alternate.

The third twig (Sycamore) also had alternate leaves; the pointed buds must have been under the leafstalks, as the leaf-scars show as rings around the buds. The larger branch grew three years ago. From the specimen one judges that the Sycamore is quite an irregularly formed tree. The twig had 11 leaves last year.

The fourth twig (Silver Maple) shows that the plant had opposite leaves, and supernumerary buds at the sides of the true axillary ones; the true axillary buds are smaller than those at the sides. It would, in such cases, be reasonable to suppose that the supernumerary buds were floral ones, and that the plant blooms before the leaves expand. The annual growths are quite extended; two years and a part of the third make up the entire twig. If it was cut during the winter of 1891-92, it must have had leaves on the lower part in 1889 and 12 leaves on the middle portion in 1890, as well as probably 4 on the lower portion on the side shoots. Last year it had 14 leaves on the end portion, two at least on each side shoot below, making 24 in all.

Folding of Leaves in the Bud.

There are some peculiarities in the arrangement of leaves in the bud which can be investigated only in the early spring. The common plans among trees are—*Inflexed*: blade folded crosswise, thus bringing it upon the footstalk. Tulip-tree. *Conduplicate*: blade folded along [Pg 34] the midrib, bringing the two halves together. Peach. *Plicate*: folded several times lengthwise, like a fan. Birch. *Convolute*: rolled edgewise from one edge to the other. Plum. *Involute*: both edges rolled in toward the midrib on the upper side. Apple. *Revolute*: both edges rolled backward. Willow. *Obvolute*: folded together, but the opposite leaves half inclosing each other. Dogwood.

Chapter VI. [Pg 35]
The Preparation of a Collection.

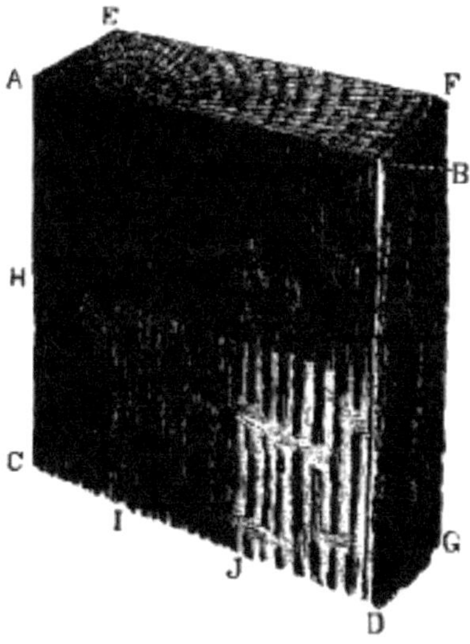

Fig. 16.

Three specimens are needed of each kind of tree: one, a branch showing the flowers; another, showing the fruit—one of these, and in many cases both, will show the leaves. The third specimen, cut from a large limb or trunk, shows the bark and the wood. This should be a specimen with a surface so cut as to show the wood in the direction of the silver grain, *radial section*; with another surface cut in the direction of the annual layers, *tangential section*; and with a third cut across the grain, *cross-section*. It should be a specimen old enough to show the change of color in the heart-wood. By taking a limb or trunk 8 inches in diameter, all these points can be secured. A specimen cut as shown in the figure will illustrate all the desired points. Side E F G shows sap-and heart-wood in tangential section; side A B D C shows the same in radial section; end A B F E, in cross-section; [Pg 36] and B F G D shows the bark. The central pith is at I; the heart-wood extends from C to J; the sap-wood from J to D. The silver grain is well shown at the end, and the blotches formed by it on the radial section.

By having the piece made smooth, and the upper part down to the center (H) varnished, the appearance of the wood in furniture or inside finish will be illustrated.

The specimens should be as nearly uniform in size as possible. If a limb 8 inches in diameter be taken and a length of 6 inches be cut off, the section A B D C should pass through the line of pith; the section E F G should be parallel with this at a distance from it of two inches; and two inches from the line of pith, the section A E C should be made. The whole specimen will then be 6 inches wide and long, and 2 inches thick.

The twigs containing leaves, flowers and fruit need to be pressed while drying in order that they may be kept in good form and made tough enough to be retained as specimens. The plants should be placed between a large supply of newspapers, or, better still, un-tarred building-felt, while drying. A weight of from 40 to 80 pounds is needed to produce the requisite pressure. The weight is placed upon a board covering the pile of plants and paper. On account of the size of many leaves and flower-clusters, these pressed speci-mens of trees should not be shorter than from 12 to 15 inches, and even a length of 18 inches is an advantage. The pads or newspapers should be about 12 by 18 inches. A transfer of the plants into dry pads each day for a few days will hasten the drying and increase the beauty of the specimens. The specimens of twigs can be mounted on cardboard by being partly pasted and partly secured by narrow strips of gummed cloth placed across the heavier portions. The cardboard should be uniform in size. One of the regular sizes of Bristol-board is 22 by 28 inches; this will cut into four pieces 11 by 14. Specimens not over 15 inches in length can readily be mounted on these, and for most collectors this might be [Pg 37] a very con-venient size. Another regular size is 22 by 32 inches, cutting well into pieces 11 by 16. Specimens 15 to 18 inches long can be mounted on these.

Some kinds of Evergreens, the Spruces especially, tend to shed their leaves after pressing. Such kinds can in most cases be made to form good specimens without pressing. Fasten the fresh specimens on pillars of plaster in boxes or frames 2 to 3 inches deep, so that they touch nothing but the column of plaster. Mix calcined plaster in water (as plasterers do), and build up a column high enough to support the branch. Place the specimen on the top of the pillar already formed, and pour over the whole some quite thin plaster till a rounded top is formed completely fastening the specimen. If the leaves are not touched at all, after they are dry, they will hang on for a long time, making specimens that will show the tree characteristics better than pressed specimens possibly could.

Chapter VII. [Pg 38]
Figures to be used in Botanical Description.

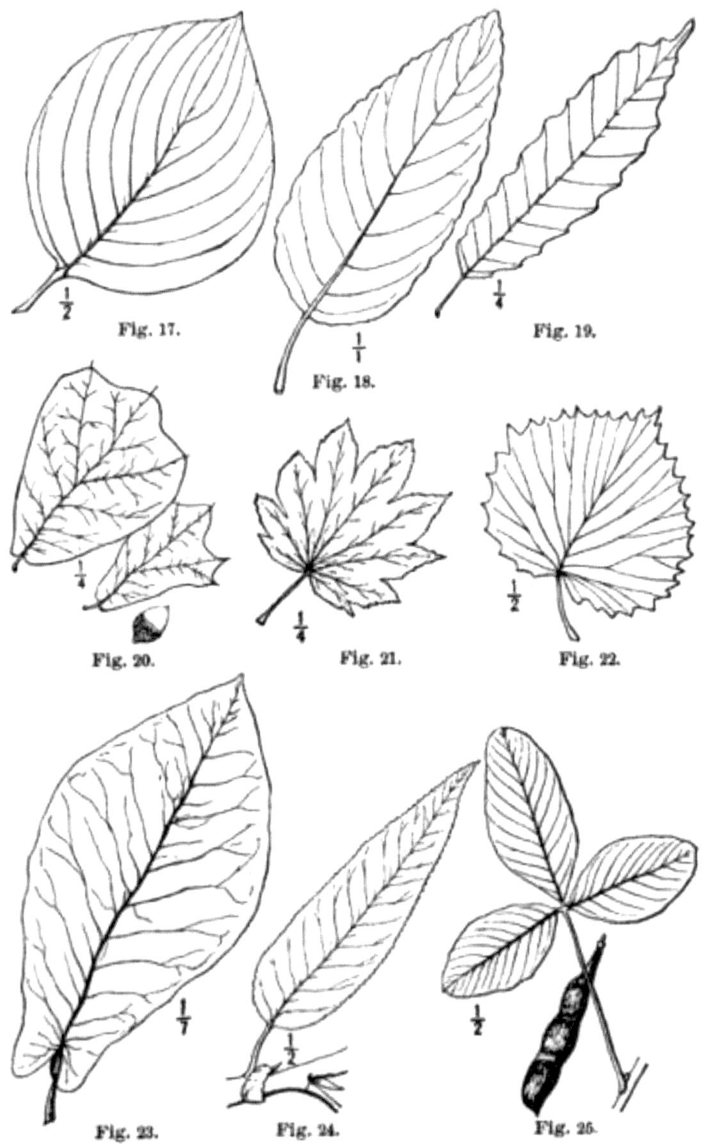

Fig. 17-25.

[Pg 39]

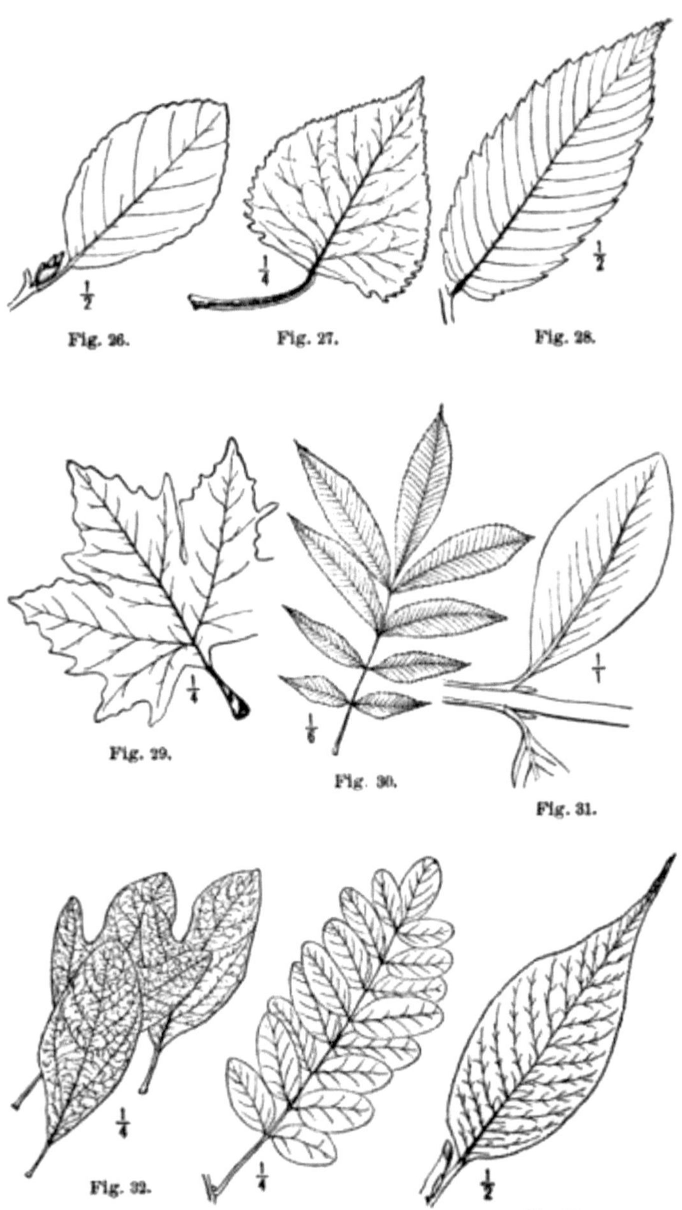

Fig. 26.   Fig. 27.   Fig. 28.

Fig. 29.

Fig. 30.

Fig. 31.

Fig. 32.   Fig. 33.   Fig. 34.

Fig. 26-34.

[Pg 40]

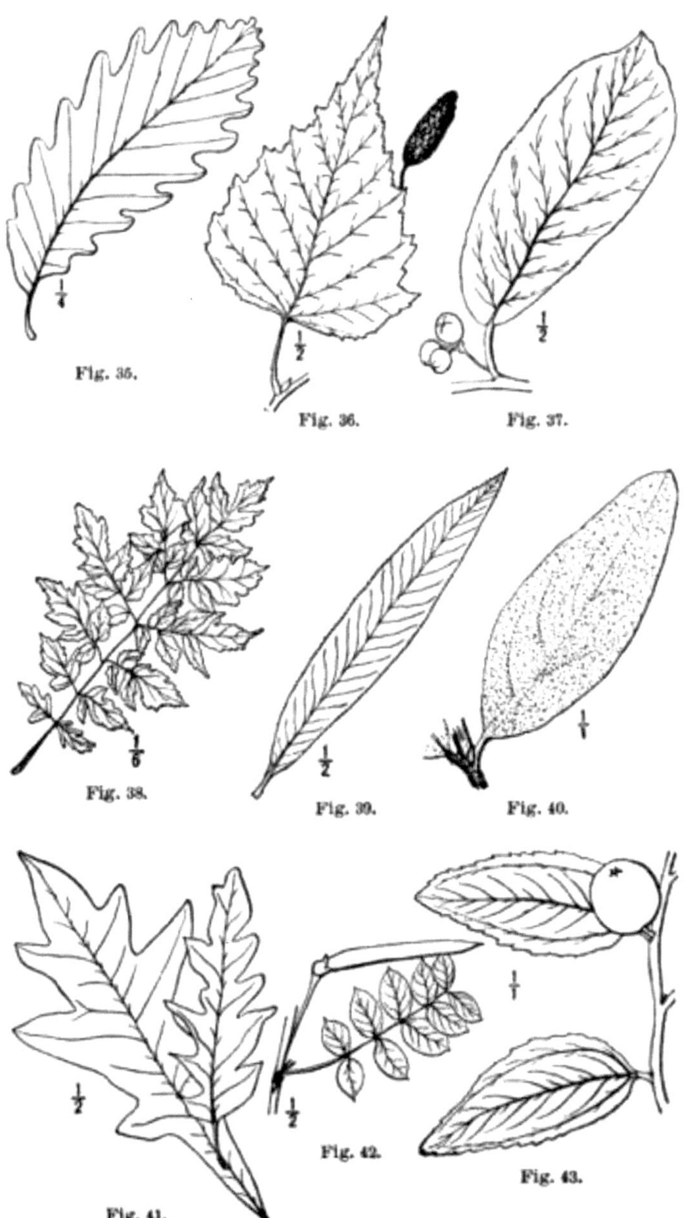

Fig. 35.

Fig. 36.

Fig. 37.

Fig. 38.

Fig. 39.

Fig. 40.

Fig. 41.

Fig. 42.

Fig. 43.

44

Fig. 35-43.

[Pg 41]

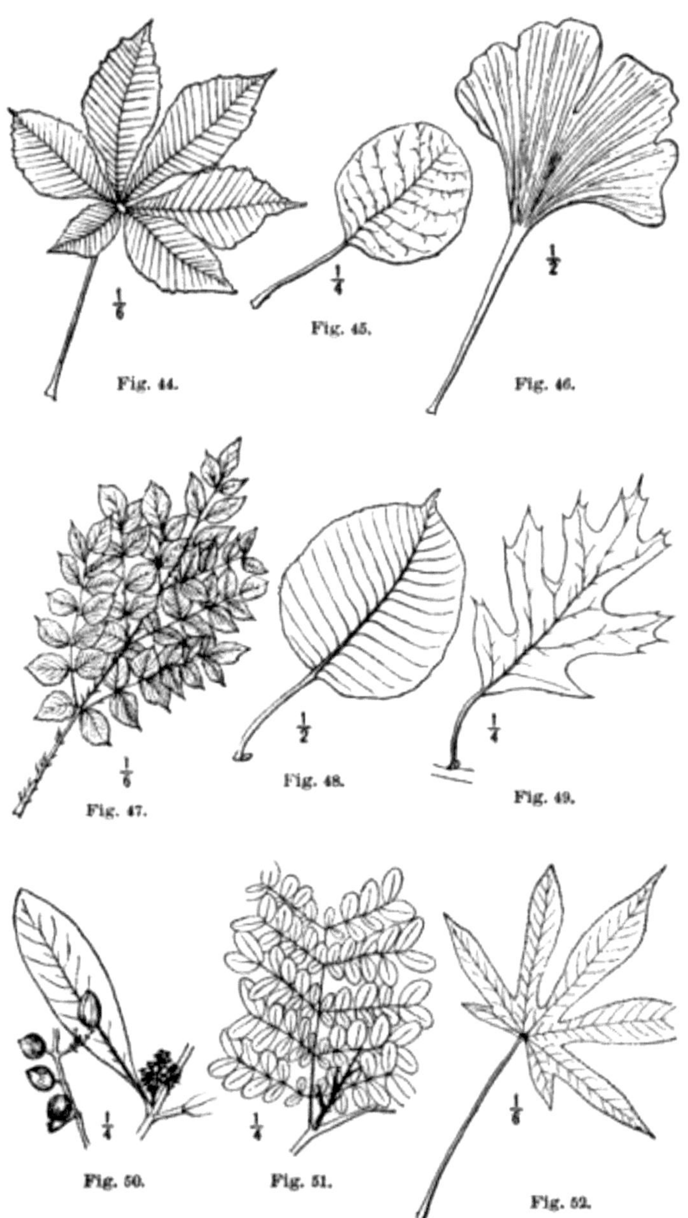

Fig. 44.

Fig. 45.

Fig. 46.

Fig. 47.

Fig. 48.

Fig. 49.

Fig. 50.

Fig. 51.

Fig. 52.

Fig. 44-52.

[Pg 42]

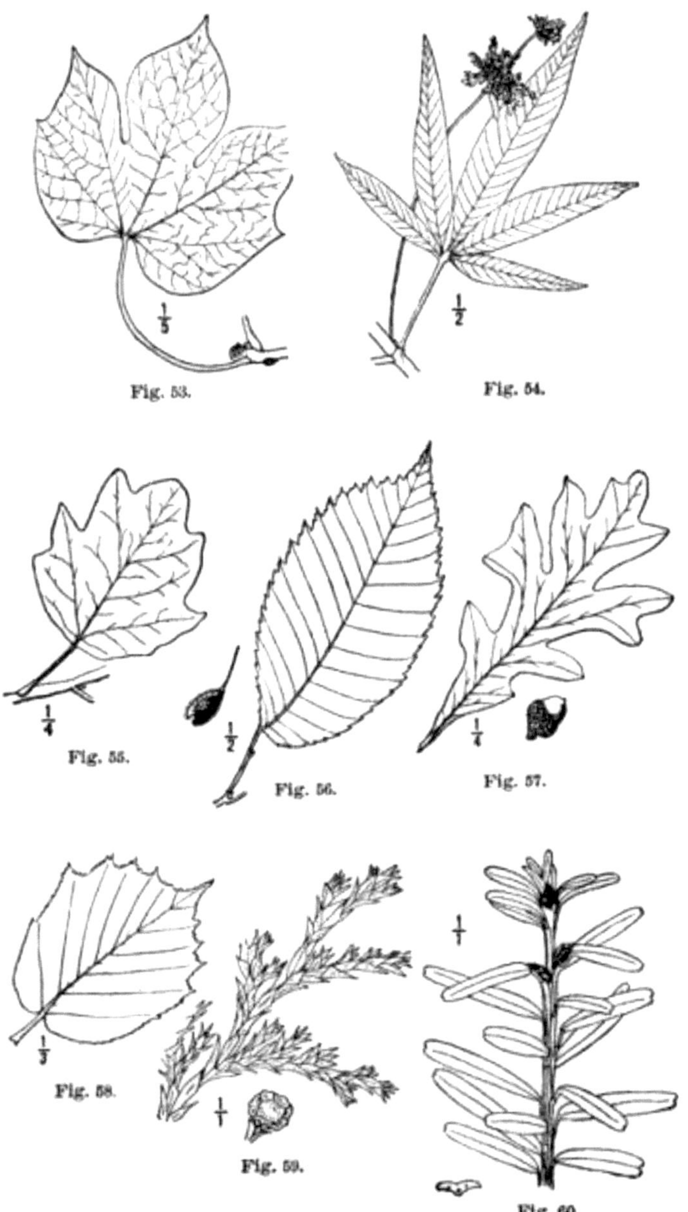

Fig. 53.

Fig. 54.

Fig. 55.

Fig. 56.

Fig. 57.

Fig. 58.

Fig. 59.

Fig. 60.

Fig. 53-60.

[Pg 43]

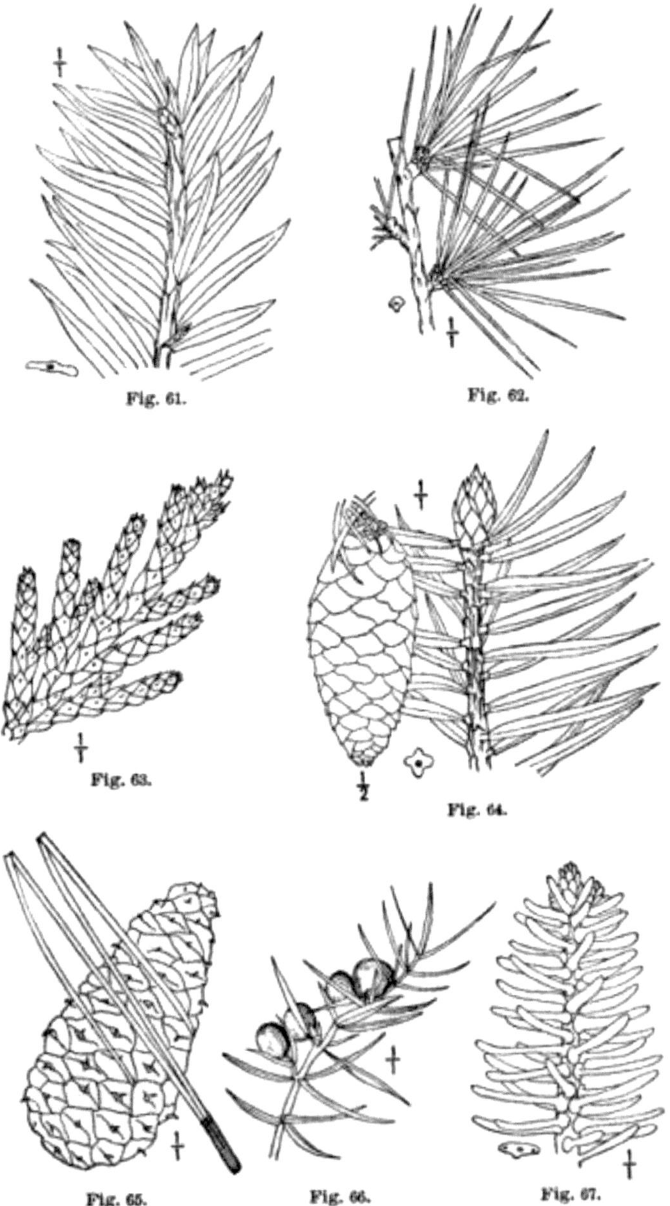

Fig. 61.

Fig. 62.

Fig. 63.

Fig. 64.

Fig. 65.

Fig. 66.

Fig. 67.

Fig. 61-67.

PART II. [Pg 44]
PLAN AND MODELS FOR TREE DESCRIPTION

All pupils should be required to write some form of composition on the trees of the region. As far as possible, these compositions should be the result of personal investigation. It is not what a pupil can read and redescribe in more or less his own words, but how accurately he can see and, from the information conveyed by his own senses, describe in his own way the things he has observed, that makes the use of such a book as this important as an educational aid. Some information in regard to trees, in a finished description, must be obtained from books, such as hardiness, geographical distribution, etc. Pupils generally should be required to include only those things which they can give from actual observation.

There are four distinct forms of tree descriptions that might be recognized by the teacher and occasionally called for as work from the pupil. 1st. A bare skeleton description, written by aid of a topical outline, from the observation of a single tree and its parts. 2d. A connected description, conveying as many facts given in the outline as can well be brought into good English sentences. This again is the description of a single tree. 3d. A connected, readable description of a certain kind of tree, made up from the observation of many trees of the same species to be found in the neighborhood. 4th. The third description including information to be obtained from outside sources in regard to the origin, geographical distribution, hardiness, character of wood, habits, durability, etc. These four plans of description [Pg 45] are more or less successive methods to be introduced as the work of a class. Pupils should be induced to carry on their own investigations as far as possible before going to printed sources for information. A good part of class work should be devoted to the first three of the methods given, but the work might finally include the fourth form of composition. The first two methods should follow each other with each of the trees studied; that is, one week let a mere outline be written, to be followed the next week with as clear and connected a description as the ability of the pupil will allow, and containing as much of the information given in the outline as possible.

Outline for Tree Description.

The tree as a whole: size, general form, trunk, branching, twigs, character of bark, color of bark on trunk, branches, and fine spray.

Leaves: parts, arrangement, kinds, size, thickness, form, edges, veining, color, surface, duration.

Buds: position, size, form, covering, number, color.

Sap and *juice*.

Flowers: size, shape, color, parts, odor, position, time of blooming, duration.

Fruit: size, kind, form, color when young and when ripe, time of ripening, substance, seeds, duration, usefulness.

Wood (often necessarily omitted): hardness, weight, color, grain, markings, durability.

Remarks: the peculiarities not brought out by the above outline.

Notes on the Foregoing Outline.

The height of a tree can be readily determined by the following plan. Measure the height you can easily reach from the ground in feet and inches. Step to the trunk [Pg 46] of the tree you wish to measure and, reaching up to this height, pin a piece of white paper on the tree. Step back a distance equal to three or four times the height of the tree; hold a lead-pencil upright between the thumb and forefinger at arm's-length. Fix it so that the end of the pencil shall be in line with the paper on the trunk; move the thumb down the pencil till it is in line with the ground at the base of the tree; move the arm and pencil upward till the thumb is in line with the paper, and note where the end of the pencil comes on the tree. Again move the pencil till the thumb is in line with the new position, and so continue the process till the top of the tree is reached. The number of the measures multiplied by the height you can reach will give quite accurately the height of the tree.

The width of the tree can be determined in the same manner, the pencil, however, being held horizontally.

In giving the forms of trees, it is well to accompany the description with a penciled outline.

The distance from the ground at which the trunk begins to branch and the extent of the branching should be noted. The direction taken by the branches, as well as the regularity and the irregularity of their position, should also be observed and described.

Although most twigs are cylindrical, still there are enough exceptions to make it necessary to examine them with reference to their form.

Under leaves, it will be well to make drawings, both of the outline and of the veining.

Crushed leaves will give the odor, and the sap can best be noticed at the bases of young leaves. The differences in sap and juice need the following words for their description: *watery, milky, mucilaginous, aromatic, spicy, sweet, gummy, resinous.*

Pupils should not always be expected to find out much about the flowers of a tree, as they are frequently very evanescent, and usually difficult to reach. [Pg 47]

The fruit lasts a greater length of time and, usually dropping spontaneously, gives a much better chance for investigation.

Specimens of most of the common woods may be obtained from cabinet-makers and carpenters. In cases where these specimens are at hand, description of the wood should be required. If the school has such specimens as are described in Chapter VI., Part I., the wood in all its peculiarities can be described.

Examples of Tree Description.

Taxodium distichum (Bald Cypress).

(Atterbury's Meadow.)

No. 1.

Tree eighty-four feet tall, thirty feet wide near base, ovate, conical, pointed; trunk seven feet in circumference near base and ridged lengthwise, but only four feet at the height of six feet from the ground, where it becomes round or nearly so, then gradually tapering to the top; branches small, very numerous, beginning six feet from the ground, sloping upward from the trunk at an angle of nearly forty-five degrees; twigs very slender, numerous, pendulous, two, three or even more growing together from supernumerary buds around the old scars; bark brownish, quite rough, thick and soft on the trunk, smoother on the branches, greenish on the young spray.

Leaves about sessile, without stipules, alternate, crowded, two-ranked, thin, linear, entire, parallel-veined, with midrib, dark green, smooth, deciduous. [Pg 48]

Buds show in the axils of only a few of the leaves, and are very small; but there are several supernumerary buds around many of the clusters of the shoots of the year.

Sap clear and slightly sticky with resin.

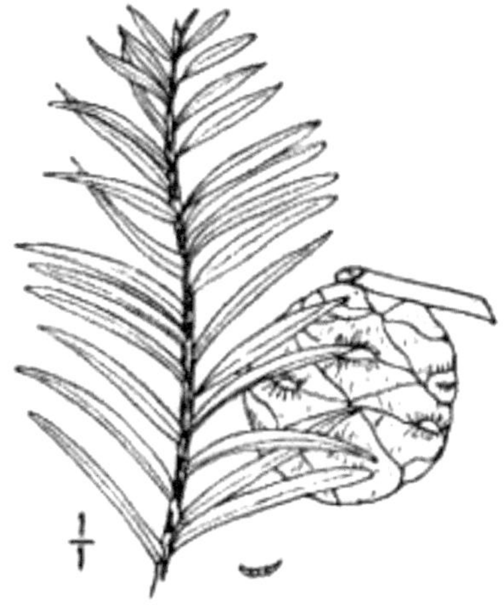

Flowers looked for, but not seen; must have been small, or have bloomed before my examination in the spring.

Fruit one inch in diameter, cone globular, brown in the autumn; did not notice it before; fifteen six-sided scales, two seeds under each, still hanging on, though the leaves have dropped; only to produce seeds, I think.

The wood I do not know about.

Remarks. Around the base, at some distance from the trunk, there are four peculiar knobs, seemingly coming from the roots, one being nearly a foot high and nine inches through.

No. 2.

The Bald Cypress standing near a small ditch in Atterbury's meadow is a very beautiful, tall, conical tree, over 80 feet high, with an excurrent trunk which is very large and ridged near the ground. It tapers rapidly upward, so that the circumference is only about half as great at the height of 6 feet, where the branches begin. The branches are very numerous and, considering the size of the trunk, very small; the largest of them being only about 2 inches through. They all slope upward rapidly, but the tip and fine spray show a tendency to droop; the fine thread-like branchlets, bearing the leaves of the year, are almost all pendulous.

The bark is very rough, thick and soft, as I found in pinning on the bit of paper to measure the height of the tree, when I could easily press the pin in to its head. [Pg 49]

The leaves are very small and delicate, and as they extend out in two ranks from the thread-like twigs, look much like fine ferns. The small linear leaves and the spray drop off together in the autumn, as I can find much of last year's foliage on the ground still fastened to the twigs. I could not see any flowers, though I looked from early in the spring till the middle of the summer; then I saw a few of the globular green cones, almost an inch in diameter, showing that it had bloomed. Next spring I shall begin to look for the blossoms before the leaves come out.

On the ground, about 6 feet from the tree, there are four very strange knobs which I did not notice till I stumbled over one of them. They seem to grow from the roots, and are quite soft and reddish in color.

No. 3.

I have found twenty-two Bald Cypresses in Trenton; they are all beautiful conical trees, and seem to grow well in almost any soil, as I have found some in very wet places and some in dry, sandy soil. They look from their position as though they had been planted out, and as I have found none in the woods around the town, they are probably not native in this region. They are from 50 to nearly 100 feet tall. I found one 96 feet high. They are all of a very symmetrical, conical form, and pointed at the top; in no case has the trunk divided into branches, and on the old trees the trunk enlarges curiously near the ground, the lower portion being very rough with ridges. The bark is very thick and rough, and is so soft that a pin can readily be pushed through it to the wood. The branches are very numerous and small, and are not regularly arranged in whorls like most of the narrow-leaved trees. These branches all slope upward from the trunk, the ends having a tendency to bend downward and make delicate drooping spray, with very small, linear, entire [Pg 50] leaves only ½ inch long. Four of the largest trees show fruit, and each of these has only about a half-dozen of the globular cones. Only a few of the trees—those in the wettest places—have the knobs on the ground near the base.

No. 4.

The Bald Cypress (*Taxodium distichum*) is a common tree, a native of the Gulf States, growing very abundantly in the wettest swamps of that region. The northern limit of the tree in its wild state is said to be central Delaware and southern Illinois, but it can be successfully cultivated in the region around Boston. There are several named varieties, one with the leaves but slightly spreading from the spray, and the whole of the branches showing a decided weeping tendency, so that it is called the Weeping Cypress. The knobs from the roots, called Cypress-knees, grow very abundantly around all the trees in the southern swamps. These grow to the height of from 2 to 4 feet, and are very thick, sometimes as much as 5 feet. They are hollow, and are occasionally used for bee-hives.

It is said to be a broad, flat-topped tree, spreading its top over other trees. This seems very strange, as none of those in Trenton, N. J., show such a tendency, but are quite spire-shaped. The wood is light, soft, straight-grained, and is said to be excellent for shingles and for other purposes. It generally has a dark reddish or brownish hue. It is a large tree, growing to the height of 140 feet. The trunk is sometimes 12 feet through near the ground. The flowers of the tree are in small catkins, blooming before the leaves expand in the early spring; in February, in South Carolina.

PART III. [Pg 51]
KEY, CLASSIFICATION AND DESCRIPTION OF THE SPECIES.

Method of Using the Key.

First read *all* the statements following the stars (*) at the beginning of the Key; decide which one of the statements best suits the specimen you have. At the end of the chosen one there is a letter in parenthesis ( ). Somewhere below, this letter is used two or more times. Read carefully *all* the statements following this letter; at the end of the one which most nearly states the facts about your specimen, you will again be directed by a letter to another part of the Key. Continue this process till, instead of a letter, there is a number and name. The name is that of the genus, and forms the first part of the scientific name of the plant. Turn to the descriptive part of the book, where this number, in regular order, is found. Here descriptions of the species of the genus are given. If there are many species, another Key will lead to the species. While the illustrations are intended to represent characteristic specimens, too much dependence must not be placed upon them; the leaves even of the same plant vary considerably, and the different varieties, especially of a cultivated plant, vary widely. Read the whole description before deciding.

The fractions beside the figures indicate the scale of the drawing as compared with the natural size of the part: 1/1 indicates natural size; 2/1, that the drawing is twice the length of the object; ¼, that the drawing is one fourth the length of the object, etc. [Pg 52]

In the description of leaves the dimensions given refer to the blade.

KEY TO THE GENERA OF TREES.

* Leaves narrow linear, needle, scale or awl shaped,
usually but not always evergreen. (**GG.**) page 60.

* Leaves broad, flat, usually deciduous, occasionally evergreen, rarely over 5 times as long as wide.
(**A.**)

A. Leaves alternate, [1] simple. (**B.**)

A. Leaves alternate, compound. (**m.**) page 57.

A. Leaves opposite or whorled on the stem. (**u.**) page 58.

B. Leaves with a midrib, netted-veined. (**C.**)

B. Leaves without a midrib, parallel-veined   109. *Salisburia.*

   C. With radiating ribs, and including those which have the lower ribs longer and more branching than those above them. (**f.**) page 56.

   C. With distinct and definite feather-veining. (**D.**)

D. Margin entire, or so nearly so as to appear entire, sometimes slightly angulated but not lobed. (**V.**)

D. Once or twice serrate or crenate or wavy-edged, but not lobed. (**E.**)

D. Distinctly lobed. (**S.**) (If the notches are over 10 on a side, look under **E.**)

   E. Straight-veined. (**M.**)

   E. Not distinctly and evenly straight-veined. (**F.**)

F. Leaves evergreen with either revolute or spiny-tipped margins   18. *Ilex.*

F. Leaves evergreen, lanceolate-oblong, minutely serrate; flowers white, 4 in. in diameter   8. *Gordonia.*

F. Leaves deciduous. (**G.**)

   G. Fruit with fleshy and often edible pulp. (**K.**)

   G. Fruit a dry and more or less rounded pod. (**H.**)

   G. Fruit and flowers in dry catkins; leaves, in most species, 3 or more times as long as wide, finely serrate to entire, with free stipules, in many species remaining on the young twigs, in others shown by a rounded scar on the sides of the stem; wood soft; the Willows   91. *Salix.*

   G. Fruit dry akenes with silky pappus, in small heads; whole plant whitened with scurf; leaves   49. *Baccharis.*

broadened and coarsely notched near tip; a broad spreading bush

[Pg 53] **H.** Flowers conspicuous, 1 in. or more in size, white. (**J.**)

H. Flowers quite small. (**I.**)

  I. Flowers and fruit in large panicles; leaves elongated, peach-like in shape, sour — 50. *Oxydendrum.*

  I. Flowers in terminal, erect racemes; fruit small, three-celled pods; leaves oval, 3-7 in. long, pointed, thin, finely serrate; plant hardly a tree — 53. *Clethra.*

  I. Fruit rounded, small, with calyx adhering to the lower part, one-seeded, in clusters of 3-many; leaves 1-3 in. long. — 56. *Styrax.*

  I. Fruit hairy, in long, hanging panicles, tipped with long, persistent style, one-seeded — 57. *Pterostyrax.*

J. Flowers bell-shaped, 1 in. long; leaves widest below the middle; fruit winged pods — 58. *Halesia.*

J. Flowers spreading, 2 in. broad; leaves about twice as long as wide, widest near the center — 7. *Stuartia.*

J. Flowers spreading, 3 in. broad; leaves about 3 times as long as wide, widest near tip — 8. *Gordonia.*

  K. Fruit a plum-like drupe with a single bony stone; plant sometimes thorny — 36. *Prunus.*

  K. Fruit berry-like, ending in a conspicuous spreading calyx; plant generally quite thorny — 38. *Cratægus.*

  K. Fruit berry-like, black when ripe, small, without calyx, with usually 3 cartilaginous coated seeds — 20. *Rhamnus.*

  K. Fruit berry-like, red when ripe, small, without calyx, with usually 4-6 hard-coated, grooved nutlets — 18. *Ilex.*

  K. Fruit a small or large apple-like pome, with the seeds in horny cells. (**L.**)

L. Fruit about ½ in. in diameter, sweet, in drooping racemes — 39. *Amelanchier.*

L. Fruit either sour or much larger, and not in elongated racemes

37. *Pyrus.*

   M. Leaves harsh to the touch; somewhat oblique at base; quite distinctly two-ranked; large trees

74. *Ulmus.*

   M. Leaves decidedly oblique at base; margin wavy; small tree, usually a shrub

40. *Hamamelis.*

   M. Fruit berry-like, ending in a conspicuous spreading calyx; plant generally quite thorny

38. *Cratægus.*

   M. Leaves not regularly oblique at base; plant not thorny. (**N.**)

N. Leaves thin and light, not harsh to the touch; spray light; bark smooth, in two species somewhat rough on the trunk. (**Q.**)

N. Leaves thick; edge wavy, almost lobed; fruit an acorn.

88. *Quercus.*

[Pg 54] **N.** Leaves broad for the length, generally doubly serrate or wavy and serrate; shrubs, rarely tall enough for trees. (**P.**)

N. Not included in the above. (**O.**)

   O. Leaves 3 or more times as long as wide, widest near the center; fruit a round, prickly bur with 1-3 horny-coated nuts

89. *Castanea.*

   O. Leaves widest near the sharply serrate tip, narrow and entire near the base; fruit small pods in terminal racemes; small tree or shrub

53. *Clethra.*

   O. Leaves widest near the base, usually small; bark scaling off like the Buttonwood; fruit axillary, solitary, small (¼ in.) roundish, dry drupes. A cultivated species, has rather large leaves, widest near the center

75. *Planera.*

P. Fruit an open oval woody catkin or cone, remaining on the plant through the winter

84. *Alnus.*

P. Fruit a rounded stony nut, in green leafy edged bracts; shrubs or small trees

85. *Corylus.*

   Q. Usually aromatic; bark dotted on the spray

83. *Betula.*

and with horizontal marks on the trunk, peeling off in thin, often papery layers

Q. Bark not peeling off in thin layers. (**R.**)

R. Leaf-buds long and slender; fruit a small prickly bur with two triangular, horny-coated nuts; large trees — 90. *Fagus.*

R. Fruit an elongated catkin with large leaf-like bracts; bark close, gray, on a grooved trunk — 87. *Carpinus.*

R. Fruit a hop-like catkin; bark brownish, finely furrowed — 86. *Ostrya.*

S. Plant more or less thorny; shrub or small tree; fruit rounded berries ending in persistent calyx-lobes — 38. *Cratægus.*

S. Plant not thorny. (**T.**)

T. Leaf deeply pinnatifid, usually with the basal lobes completely separated; cultivated — 37. *Pyrus.*

T. End of leaf as though cut off; sides with one large lobe; margin entire; large tree — 2. *Liriodendron.*

T. Lower leaves three-lobed, heart-shaped at base, upper merely ovate, margin entire; small tree or shrub — 66. *Clerodendron.*

T. Not as above; leaves usually many-lobed. (**U.**)

U. Leaves thin; bark of trunk peeling off in thin horizontal strips — 83. *Betula.*

U. Leaves thin; leaf-buds long, slender, sharp-pointed; bark smooth, not peeling; cultivated — 90. *Fagus.*

U. Leaves thickish; bark roughish; fruit an oval woody cone, remaining on through the year — 84. *Alnus.*

U. Leaves thick; fruit an acorn — 88. *Quercus.*

[Pg 55] **V.** Leaves evergreen, small, 2-3 in. long, thick, with revolute margins; fruit an acorn — 88. *Quercus.*

V. Leaves evergreen, oval to lance-oval, usually large; small trees, almost shrubs. (**d.**) page 56.

V. Leaves deciduous (some are evergreen in the

Southern States). (**W.**)

   W. Plant more or less spiny. (**c.**)

   W. Plant not at all spiny. (**X.**)

X. Leaf-blade thin, long, pointed, with curved parallel veins or ribs   45. *Cornus.*

X. Leaf-blade thin, circular or broadly oval in outline, with blunt, almost rounded apex; veins not regularly parallel   27. *Rhus.*

X. Leaf quite elongated, 5 or more times as long as wide. (**b.**)

X. Leaves with none of the above peculiarities. (**Y.**)

   Y. Deciduous bud-scales (stipules), leaving a scar or mark completely around the stem at the base of the leaves.   1. *Magnolia.*

   Y. Leaves covered on one or both sides with silvery scales   71. *Elæagnus.*

   Y. No such ring around the stem, or silvery scales on the leaves. (**Z.**)

Z. Leaves distinctly straight-veined, thin   90. *Fagus.*

Z. Leaves thick, obtuse; fruit an acorn   88. *Quercus.*

Z. Leaves 6 in. or more long; crushed leaves with a rank, fetid odor   5. *Asimina.*

Z. Leaves 3-5 in. long; twigs and leaves very spicy; shrub rather than tree   70. *Lindera.*

Z. Leaves about 2 in. long, oval, on twigs which have ridges extending down from the sides of the leafstalk; small tree, almost a shrub, with beautiful flowers   43. *Lagerstroemia.*

Z. Leaves not as above. (**a.**)

   a. Fruit a large (½-1½ in.) rounded pulpy berry with a heavy calyx at the base   55. *Diospyros.*

   a. Fruit small (¼ in.), fleshy, drupe-like, with a striate stone; limbs branching horizontally, often descending   46. *Nyssa.*

a. Fruit a black, juicy berry (1/3-½ in.), with about 3 seeds   20. *Rhamnus.*

a. Fruit an ovoid dry drupe (½ in.); leaves sweet-tasting   59. *Symplocos.*

a. Fruit an apple-like pome (Quince)   37. *Pyrus.*

b. Wood soft; both kinds of flowers in catkins in spring; with either stipules or stipular sears   91. *Salix.*

b. Wood hard; leaves thick; fruit an acorn   88. *Quercus.*

c. Fruit a 2-4-seeded small berry; juice not milky   20. *Rhamnus.*

[Pg 56] **c.** Fruit large, orange-like in size and color when ripe; juice milky   77. *Maclura.*

c. Fruit small, black when ripe, cherry-like; juice milky   54. *Bumelia.*

d. Aromatic; berries dark blue on red stalks   68. *Persea.*

d. Not aromatic; leaves nearly 1 ft. long; flowers large and solitary.   1. *Magnolia.*

d. Not aromatic; leaves 1-4 in. long; flowers very small; fruit small dark-colored berries, with 2-4 seeds   20. *Rhamnus.*

d. Not aromatic; flowers large, in showy clusters. (**e.**)

  e. Leaves 5 in. or more long   52. *Rhododend-ron.*

  e. Leaves less than 4 in. long   51. *Kalmia.*

f. Leaves decidedly aromatic, usually somewhat irregularly lobed, margin entire, base tapering   69. *Sassafras.*

f. Leaves usually deltoid, sometimes heart-shaped with serrate margin and gummy buds, rarely palmately lobed. All have either the petiole flattened sidewise, the leaf-blade densely silvery-white beneath, or gummy aromatic buds   92. *Populus.*

f. Leaves broadly heart-shaped; margin entire; small tree with abundance of red flowers in early spring; fruit a pea-like pod.   32. *Cercis.*

f. Leaves not as above given. (**g.**)

  g. Leaves broadly heart-shaped, with a serrate margin and a petiole about as long as the blade, sometimes longer; base of leaf not oblique   4. *Idesia.*

  g. Leaves broadly heart-shaped, those on the suckers much lobed; base not oblique; margin serrate; juice milky; bark very tough. (**l.**)

  g. Leaves broadly heart-shaped, with an oblique base; margin regularly serrate; juice not milky   11. *Tilia.*

  g. Leaves slightly if at all heart-shaped at base, usually somewhat oblique, with neither milky juice nor lobes. (**j.**)

  g. Leaves decidedly and quite regularly lobed. (**h.**)

h. Leaves with 3-5 large lobes, the margin entire or slightly angulated.   10. *Sterculia.*

h. Leaves star-shaped, with 5-9 pointed, serrate lobes. (**i.**)

h. Leaves large, irregularly margined; leaf-stem covering the bud; large tree   80. *Platanus.*

h. Plant quite thorny; fruit berry-like, ending in a conspicuous spreading calyx; small trees or shrubs with apple-like blossoms.   38. *Cratægus.*

h. Leaves with a tapering base; small tree, almost a shrub, with large Hollyhock-like flowers; plant not thorny   9. *Hibiscus.*

  [Pg 57] **i.** Large tree, with fruit 1 in. in diameter, dry, rough, hanging on a long stem   41. *Liquidambar.*

  i. Small tree with few branches and the trunk usually quite prickly; fruit berry-like in large clusters   44. *Aralia.*

j. Fruit small berries, with 3 flattened seeds, in clusters in the axils of the leaves, which are decidedly 3-ribbed from the base   21. *Hovenia.*

j. Fruit small drupes, with 1 seed, either solitary or

in pairs in the axils of the leaves. (**k.**)

   k. Plant without prickles; leaves decidedly oblique at base   76. *Celtis.*

   k. Plant with prickles; leaves narrow, decidedly 3-ribbed, and 2-ranked on green twigs   22. *Zizyphus.*

l. Fruit not very edible; leaves rough above, very hairy below, on some of the twigs opposite   79. *Brousso-netia.*

l. Fruit edible; leaves not very hairy, never opposite   78. *Morus.*

   m. Leaves of 3 entire-edged leaflets; fruit a pea-like pod   28. *Laburnum.*

   m. Leaves of 3 quite regularly serrate, transparent-dotted leaflets   13. *Ptelea.*

   m. Leaves once or twice pinnate; the leaflets entire. (**s.**)

   m. Leaves once or twice pinnate; the leaflets with margins more or less serrate or notched. (**n.**)

n. Leaves irregularly once to twice, in one case three times, pinnate. (**r.**)

n. Leaves regularly once pinnate. (**o.**)

   o. Leaves less than 1 ft. long, on a small, quite prickly plant; fruit very small pods (¼ in. long)   12. *Xan-thoxylum.*

   o. Leaves less than 1 ft. long; leaflets 3 in. or less long; fruit bright-colored, berry-like pomes in clusters, persistent through the autumn; plant not thorny; branches not heavy-tipped.   37. *Pyrus.*

   o. Leaves usually larger on the small tree or almost a shrub; juice in most cases milky; branches heavy-tipped   27. *Rhus.*

   o. Leaves 1-2 ft. long; leaflets 3 in. or more long; fruit a bony nut with green fleshy coat; large trees. (**q.**)

   o. Leaves very large, 2 ft. or more long on the rapid-growing branches; branches heavy-tipped; odor of bruised leaves quite strong; leaflets 15 or more in number; large trees; juice not milky. (**p.**)

p. Leaflets with 1-3 glandular notches at the base   17. *Ailanthus.*

p. Leaflets entire at base, but very slightly serrate near the tip   16. *Cedrela.*

  [Pg 58] **q.** Coat of fruit more or less dehiscent into 4 valves; nut smoothish; leaflets, except in one species, not over 11 in number, usually 5-7   82. *Carya.*

  q. Coat of fruit not regularly dehiscent; nut, in the wild species, rough-coated; leaflets, except in a cultivated species, over 11 in number   81. *Juglans.*

r. Leaves quite regularly twice odd-pinnate; leaflets about 1 in. long; juice not milky; fruit rounded berries in large clusters; plant not prickly; branchlets not heavy-tipped   15. *Melia.*

r. Leaves once to twice irregularly odd-pinnate; the leaflets very irregularly and coarsely toothed; a small, round-headed tree with bladdery pods   24. *Koelreuteria.*

r. Leaves irregularly about twice odd-pinnate; the leaflets lanceolate; quite a low plant with few heavy-tipped branches; plant without prickles   27. *Rhus.*

r. Leaves 2 (sometimes 3) times odd-pinnate; tree-stem with prickles; small tree or shrub, with few branches   44. *Aralia.*

r. Leaves once to twice abruptly pinnate; large tree with slender-tipped branches, usually very thorny   34. *Gleditschia.*

  s. Leaves very large (2 ft. or more long), about twice abruptly pinnate; leaflets broad and often 2 in. long; branches blunt; no thorns   33. *Gymnocla-dus.*

  s. Leaves and leaflets much smaller, leaves quite irregularly once or twice abruptly pinnate; branches slender-tipped; large tree, usually very thorny   34. *Gleditschia.*

  s. Leaves twice abruptly pinnate; leaflets over 400 in number, with midrib near the upper edge   35. *Albizzia.*

  s. Leaves regularly once pinnate, not over 2 ft. long. (**t.**)

t. Leaves abruptly pinnate, not over 5 in. long; leaflets 8-12, small, mucronate-pointed — 29. *Caragana.*

t. Leaves odd-pinnate; shrub or small tree, with few, heavy-tipped branches; no spines or prickles — 27. *Rhus.*

t. Leaves odd-pinnate; leaflets large (3-5 in. long), not usually over 11 in number; round-topped tree — 30. *Cladrastis.*

t. Leaves odd-pinnate; leaflets less than 3 in. long, frequently 11-21 in number; often with spines at the bases of the leaves in the place of stipules — 12. *Xanthoxylum* or 31. *Robinia.*

  u. Leaves palmately compound. (**CC.**)

  u. Leaves pinnately compound. (**BB.**)

  u. Leaves simple, evergreen, sessile, in whorls around the stem, which they completely cover — (98a. *Araucaria.*)

  u. Leaves simple, opposite, evergreen, entire, over 2 in. long — 61. *Osmanthus.*

  [Pg 59] **u.** Leaves simple, opposite, evergreen, entire, under 1 in. long — 73. *Buxus.*

  u. Leaves simple, deciduous. (**v.**)

v. Branches ending in thorns; small trees, or shrubs. (**AA.**)

v. Plants not thorny. (**w.**)

  w. Leaves palmately lobed (one variety, rarely cultivated, lacks lobes, but is heart-shaped with a serrate margin), the lobes over 3 in number, or with notches or serrations; fruit dry, winged — 25. *Acer.*

  w. Lower leaves palmately 3-lobed, and heart-shaped at base, upper ones ovate, all with entire margin; fruit with juicy pulp covering the 4 seeds — 66. *Clerodendron.*

  w. Leaves palmately lobed; fruit small, one-seeded, berry-like drupes in large clusters, with flattened stones, or large rounded clusters of flowers without stamens or pistils; shrubs rather than trees — 47. *Viburnum.*

  w. Leaves heart-shaped, entire or slightly angulated; not lobed. (**DD.**)

w. Leaves irregularly serrate, somewhat straight-veined; fruit single-winged; large cultivated tree — 60. *Fraxinus.*

w. Leaves neither heart-shaped nor lobed; small trees, almost shrubs. (**x.**)

x. Leaves entire. (**z.**)

x. Leaves serrate or dentate, ovate or oval. (**y.**)

y. Fruit rounded drupes in large clusters, with single flattened stones — 47. *Viburnum.*

y. Fruit lobed pods, which burst open in the autumn; branchlets somewhat 4-sided — 19. *Euonymus.*

z. Leaves small, lanceolate; flowers and fruit large and beautiful — 42. *Punica.*

z. Leaves broad, thin, with curved parallel veins or ribs. — 45. *Cornus.*

z. Leaves large, broad, oval, without either curved or straight parallel ribs — 63. *Chionanthus.*

AA. Leaves entire and covered on both sides with silvery, peltate scales — 72. *Shepherdia.*

AA. Leaves ovate, small, minutely serrate — 20. *Rhamnus.*

BB. Leaves large, 18 in. or more long; leaflets 11 or more, very finely serrated — 14. *Phellodendron.*

BB. Leaves smaller; leaflets entire or quite evenly toothed, usually over 5 in number — 60. *Fraxinus.*

BB. Leaflets coarsely and quite irregularly toothed, 3-5 (rarely 7) in number — 26. *Negundo.*

CC. Leaflets slender-lanceolate, almost entire; shrub or small tree, 5-10 ft. high — 67. *Vitex.*

[Pg 60] **CC.** Leaflets broader and serrate; usually large trees. — 23. *Æsculus.*

DD. Leaves with radiating ribs. (**FF.**)

DD. Leaves with feather-veining. (**EE.**)

EE. Leaves 2-6 in. long; flowers small, in large, dense, terminal clusters — 62. *Syringa.*

EE. Leaves 1-4 in. long; flowers in pairs — 48. *Lonicera.*

FF. Leaves large, 6 in. or more long; two almost hidden buds, one above the other, in the axils of the leaves on the rapid-growing branches; flowers large, purple, blooming in early spring; fruit rounded pods — 64. *Paulownia.*

FF. Leaves large, 6 in. or more long; flowers large, white, blooming in June; fruit long pods — 65. *Catalpa.*

FF. Leaves 2-4 in. long, with red stems — 3. *Cercidiphyllum.*

   GG. Leaves scattered singly over the stem, not in bundles or clusters. (**JJ.**)

   GG. Leaves in large or small clusters. (**HH.**)

HH. Clusters in whorls of many leaves around the stem like an umbrella — 100. *Sciadopitys.*

HH. Leaves clustered in bundles of 2-6 — 93. *Pinus.*

HH. Leaves clustered in bundles of over 8. (**II.**)

   II. Leaves deciduous, soft — 97. *Larix.*

   II. Leaves evergreen, rigid — 98. *Cedrus.*

JJ. Leaves hardly evergreen; spray quite slender. (**ZZ.**)

JJ. Leaves fully evergreen. (**KK.**)

   KK. Leaves awl or scale shaped, and mainly appressed to the stem. (**WW.**)

   KK. Leaves linear or needle shaped, and decidedly spreading from the stem, though sometimes with a decurrent base. (**LL.**)

LL. Leaves narrowed to a distinct though short stem. (**RR.**)

LL. Leaves sessile; if narrowed, not so abruptly as to form a petiole. (**MM.**)

   MM. Leaves opposite or whorled on the stem. (**PP.**)

   MM. Leaves rather spirally arranged around the stem, not just opposite. (**NN.**)

NN. Leaves linear to lanceolate, flattened, spreading quite squarely from the stem. (**OO.**)

NN. Leaves not flattened but 4-sided, curved, gradually enlarging from the tips to the bases, which are decurrent, and on the young twigs completely cover the stem; cones rounded; the scales not lapping — 105. *Cryptomeria.*

    OO. Leaves about linear in form, of nearly the same width throughout, and usually fastened to the cylindrical stem by a distinct disk-like base; cones erect; scales lapping. — 96. *Abies.*

    [Pg 61] **OO.** Leaves about 2 in. long and gradually widening from the acute tips to the broad (1/8 in.) bases, which are decurrent on the stem — 99. *Cunninghamia.*

    OO. Leaves ½-1 in. long, sharp-pointed, very flat, two-ranked, somewhat lanceolate in form; base narrowed almost to a petiole — 102. *Sequoia.*

PP. Leaves not decurrent, usually in whorls of three around the stem, sometimes opposite, acute-pointed; fruit small (1/8 in.), rounded, dark-colored berries — 106. *Juniperus.*

PP. Leaves decurrent on the stem, less than ½ in. long. (**QQ.**)

    QQ. Fruit small, globular cones; the scales not lapping — 104. *Chamæcyparis.*

    QQ. Fruit small, elongated cones of few, lapping scales — 103. *Thuya.*

RR. Leaves usually but little flattened, but jointed to a short, brown petiole which is attached to a somewhat grooved twig; cones pendent, of lapping scales — 94. *Picea.*

RR. Leaves decidedly flattened, not jointed, but narrowed to a petiole which is usually green or greenish in color. (**SS.**)

    SS. Leaves rounded or obtuse at the tip, distinctly two-ranked, usually less than 1 in. long; cones — 95. *Tsuga.*

oval, 1 in. or less long, of lapping scales

SS. Leaves acute at the tip; fruit (found only on a portion of the plants, as the flowers are dioecious) drupe-like, with a single nut-like seed. (**TT.**)

TT. Leaves not two-ranked, over 2 in. long — 108. *Podocarpus.*

TT. Leaves quite regularly two-ranked. (**UU.**)

UU. Leaves marked by two longitudinal lines; bruised or burned leaves with a very disagreeable odor — (107a. *Torreya.*)

UU. Leaves with the midrib forming a distinct ridge, odor not disagreeable. (**VV.**)

VV. Leaves usually less than an inch long — 107. *Taxus.*

VV. Leaves usually more than an inch long — (107b. *Cephalotaxus.*)

WW. Spray decidedly two-ranked, fan-like. (**YY.**)

WW. Spray branching in an irregular way, not two-ranked. (**XX.**)

XX. Fruit a purplish berry; bark shreddy — 106. *Juniperus.*

XX. Fruit a cone of thick, pointed, not lapping scales — 102. *Sequoia.*

YY. Cones elongated, of lapping scales — 103. *Thuya.*

YY. Cones globular, of peltate, valvate scales — 104. *Chamæcyparis.*

[Pg 62] **ZZ.** Leaves very broad at base, half clasping the stem and rapidly narrowed to an acute tip; hardly at all spreading from the thread-like twigs; flowers pinkish, in spike-like clusters — 6. *Tamarix.*

ZZ. Leaves more elongated, quite even in width, not clasping the stem — 101. *Taxodium.*

[1] Look on the elongated branches for the arrangement of the leaves; they are too closely clustered on the short side shoots. See page 18.

CLASS I. ANGIOSPÉRMÆ.

Plants with a pistil consisting of a closed ovary, which contains the ovules and forms the fruit.

Order **I. MAGNOLIACEÆ.** (Magnolia Family.)

Trees or shrubs, mainly of tropical regions, including, in our section, the three following genera:

Genus **1. MAGNÒLIA.**

Trees and tall shrubs with alternate, thick, smooth, entire leaves with deciduous stipules which form the bud-scales, and are attached entirely around the stem, leaving a ridge, as in Liriodendron.

Flowers very large (3 to 10 in. in diameter), usually white, solitary.

Fruit a large cone from which the seeds, drupe-like, usually red, hang out on long threads during the autumn.

* Blooming with or before the opening of the leaves. (**A.**)

A. Flowers entirely white — 9, 10.

A. Flowers dark purple — 11.

A. Flowers mixed purple and white. A large number of hybrids from China and Japan.

* Blooming after the leaves expand. (**B.**)

B. Leaves evergreen, more than 8 in. long — 1.

B. Leaves evergreen, not 6 in. long — 2.

B. Leaves deciduous. (**C.**)

C. Leaves decidedly auriculate or cordate at the base. (**D.**) [Pg 63]

D. Leaves very large (1 to 3 ft. long) — 5.

D. Leaves smaller and much clustered at the tips of the flowering branches — 6.

C. Leaves not conspicuously cordate at base. (**E.**)

> **E.** Leaves clustered at the tips of the flowering branches   7.

> **E.** Leaves scattered along the branches. (**F.**)

>> **F.** Base of leaf abrupt   3, 4.

>> **F.** Base of leaf tapering. (**G.**)

>>> **G.** Leaves quite large, about 1 ft. long; a very erect growing tree   8.

>>> **G.** Leaves smaller, medium thick, glossy above   2.
>>> medium thin (5 to 10 in. long)   3.

M. grandiflòra.

1. **Magnòlia grandiflòra**, L. (Large-flowered Magnolia. Southern Evergreen Magnolia.) Leaves evergreen, thick, oval-oblong; upper surface glossy, under surface somewhat rusty. Flowers large, 6 to 10 in. wide, white, fragrant. In spring. Fruit oval, 3 to 4 in. long, ripe in October. Seeds scarlet. Splendid evergreen tree (50 to 80 ft.) in the Southern States; half hardy, and reduced to a shrub (10 to 20 ft.) when cultivated in the Middle States.

M. glaùca.

2. **Magnòlia glaùca**, L. (Sweet-Bay. Swamp-Magnolia.) Leaves quite thick, oblong-oval, obtuse, smooth and glossy above, white or rusty pubescent beneath; evergreen in the Southern States. Leaf-buds silky. Flowers globular, white, and very fragrant. June to August. Fruit about 1½ in. long, ripe in autumn. Shrub, 4 to 20 ft. high, in the swamps of the Atlantic States from Massachusetts southward. Slender tree, 15 to 30 ft. high, when cultivated in good damp soil.

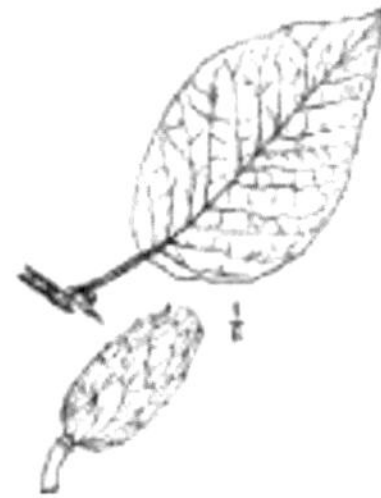

M. acuminàta.

3. **Magnòlia acuminàta**, L. (Cucumber-tree.) Leaves thin, green above, paler beneath, oblong, usually pointed at both ends, 5 to 10 in. long. Leaf-buds silky. Flowers pale yellowish-green, 3 in. wide, late in spring. Fruit irregular-oblong (2 to 3 in. long), rose [Pg 64] -colored when ripe, with a few hard, bony, black seeds, coated with red pulp, ripe in autumn. Large (50 to 90 ft.) noble forest tree, wild in western New York and southward. Wood rather soft, yellowish-white, quite durable, and extensively used for pump logs. Occasionally cultivated; fine for avenues.

M. cordàta.

4. **Magnòlia cordàta**, Michx. (Yellow Cucumber-tree.) Leaves broadly ovate or oval, rarely cordate at base, smooth above, white-downy beneath, 4 to 6 in. long. Flowers lemon-yellow slightly streaked with red. June. Fruit nearly 3 in. long, red when ripe in autumn. A rather small, broad-headed tree (20 to 50 ft.), wild in the Southern States, but hardy as far north as Boston; not often cultivated. Probably an upland variety of the preceding.

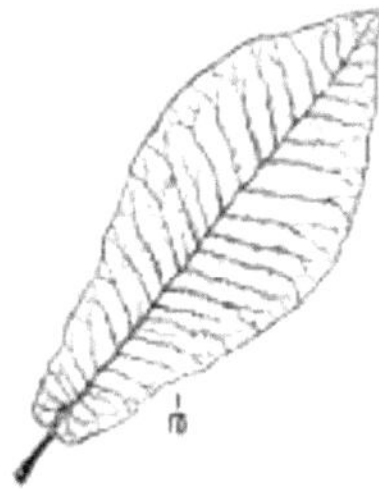

M. macrophýlla.

5. **Magnòlia macrophýlla**, Michx. (Great-leaved Magnolia.) Leaves very large, sometimes 3 ft. long, crowded at the summit of the branches, obovate-oblong, cordate at the narrowed base, glaucous-white beneath, green above; twigs whitish pubescent. Flowers very large (12 in. broad), white with a purple spot near the base; fragrant. Fruit cylindrical, 4 in. long, deep rose-colored when ripe in autumn. A medium-sized (30 to 40 ft.), spreading tree; wild from Kentucky south, hardy and cultivated as far north as New York City.

M. Fràseri.

6. **Magnòlia Fràseri**, Walt. (Ear-leaved Umbrella-tree.) Leaves crowded at the ends of the flowering branches, obovate or spatulate, auriculate at base, smooth (1 ft. long). Leaf-buds smooth. Flowers (6 in. wide) white, slightly scented. April to May. Fruit 3 to 4 in. long, rose-colored, ripe in autumn. [Pg 65] Medium-sized, rather slender tree (30 to 50 ft.), with soft yellowish-white wood. Virginia and southward. Hardy and extensively cultivated as far north as New York City.

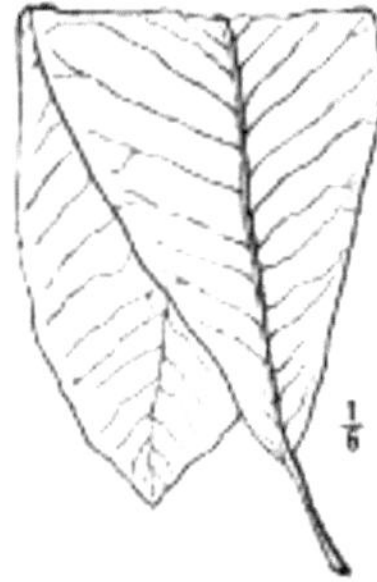

M. umbrélla.

7. **Magnòlia umbrélla**, Lam. (Umbrella tree.) Leaves clustered at the ends of the branches, obovate-lanceolate, pointed at both ends, 1 to 2 ft. long; downy beneath when young, but soon becoming smooth. Flowers white, 6 to 8 in. broad. May. Fruit oblong, 4 to 6 in. long, rather rose-colored when ripe in autumn. A small, rather straggling tree, 20 to 40 ft. high; common in the Southern States, and wild as far north as New York State; cultivated throughout.

M. hypoleùca.

8. **Magnòlia hypoleùca**, S. & Z. (Japan Magnolia.) Leaves large (1 ft. long), somewhat purple-tinted above, white and glaucous beneath. Midrib and leafstalk often red. Flowers cream-white, fragrant, appearing after the leaves in June. Twigs stout and polished. A medium-sized, very erectly growing tree; from Japan.

M. conspícua.

9. **Magnòlia conspícua,** Salisb. (Yulan or Chinese White Magnolia.) Leaves deciduous, obovate, abruptly acuminate, pubescent when young. Flowers large (4 in.), cream-white, very fragrant, appearing very early (May), before any of the leaves. Fruit rarely formed, with few (1 to 3, rarely more) seeds [Pg 66] to a cone. Bark dark brown on the young branches; terminal winter buds over ½ in. long. Small tree (10 to 30 ft.) with spreading habit and stout branches; very extensively cultivated for its abundant early bloom; from China.

M. Kòbus.

10. **Magnòlia Kòbus.** (Thurber's Japan Magnolia.) Leaves similar to the preceding, but smaller. Flowers also similar, but pure white. Fruit abundantly formed, with several (2 to 12) seeds to the cone.

Bark green on the young growth; terminal winter-buds under ½ in. long. Small tree (15 to 40 ft.) with erect habit and slender branches. A beautiful tree of recent introduction from Japan.

M. purpùrea.

11. **Magnòlia purpùrea**, Sims. (Purple Japan Magnolia.) Leaves obovate, pointed at both ends, dark green. Flowers erect, of 3 sepals and 6 obovate, purple petals; blooming about as the leaves expand. A low tree, or usually merely a shrub, from Japan; often cultivated.

Besides the Magnolias here given, there are quite a number of varieties and hybrids in cultivation, from China and Japan, most of them blooming before the leaves expand in spring.

Genus **2. LIRIODÉNDRON.**

Trees with alternate, deciduous, smooth, stipulate, 4-lobed leaves, the stipules large, attached entirely around the stem, and leaving a ridge when they drop off, as in the genus Magnolia. Flowers tulip-shaped, large (3 in.), greenish-yellow. May to June. Fruit a pointed cone, 3 in. long, hanging on the tree till autumn.

L. tulipífera.

Liriodéndron tulipífera, L. (Tulip-tree.) Leaves large, smooth on both [Pg 67] sides, somewhat 3-lobed, the end one seemingly cut off, leaving a shallow notch; stipules light-colored, large, oblong, attached all around the stem, often remaining on through half the season. A very large (80 to 150 ft. high), beautiful, rapidly growing tree, with soft, straight-grained, greenish wood, of great use for inside work. Southern New England and southward. Especially abundant and large in the Western States. Also cultivated.

Genus 3. CERCIDIPHÝLLUM.

Shrubs or trees with opposite, rarely subalternate, simple, deciduous leaves. Fruit short-stemmed, with divergent pods, 2-4 in number, splitting open on the outer edges; each one-celled, with one row of lapping, pendulous seeds with membranous wings.

C. Japónicum.

Cercidiphýllum Japónicum. (Katsura-tree.) Leaves broadly heart-shaped, palmately veined with 5-7 ribs, and with an apparently entire margin, dark green above, somewhat glaucous beneath. Under a magnifying glass the margin will be found to have pellucid crenulations. Leafstalk dark red and jointed above the base, the veins somewhat red-tinted. A beautiful, upright tree with birch-like, dotted, brown bark; of recent introduction from Japan, and probably completely hardy throughout the region.

Order II. BIXÍNEÆ.

A rather small order of mostly tropical trees or shrubs, with alternate, simple leaves.

Genus 4. IDÈSIA.

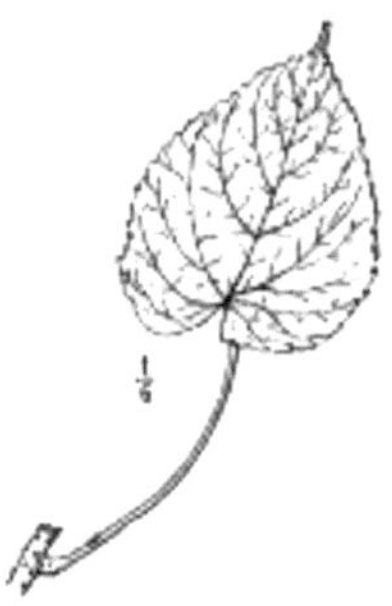

I. polycárpa.

Large trees with terminal and axillary panicles of very small flowers and berries.

Idèsia polycárpa, Hook. Leaves large, heart-shaped, serrate, palmately veined with 5 ribs; leafstalk very long, red, with two [Pg 68] glands near the base; twigs also glandular; berries very small (¼ inch), with many seeds. A large tree recently introduced from Japan, which may prove hardy from Pennsylvania south, but is killed by the climate of Massachusetts.

Order **III. ANONÀCEÆ.**

(Custard-apple Family.)

An order of tropical trees and shrubs except the following genus:

Genus **5. ASÍMINA.**

Small trees or shrubs with simple, deciduous, alternate, entire, pinnately-veined leaves. Flowers large, dull purplish, solitary in the axils of last year's leaves. Fruit a large, oblong, several-seeded, pulpy berry.

A. tríloba.

Asímina tríloba, Dunal. (Common Papaw.) Leaves large (8 to 12 in. long), oblong-obovate, acuminate, thin, lapping over each other in such a manner as to give the plant a peculiar imbricated appearance. Flowers 1 in. broad, appearing before the leaves. Fruit 3 in. long, 1½ in. thick, yellowish, fragrant, about 8-seeded, ripe in the autumn. Small (10 to 20 ft. high), beautiful tree with dark-brown twigs. All parts have a rank, fetid smell. Wild in New York and southward along streams; cultivated.

Order **IV. TAMARISCÍNEÆ.**

A small order, consisting mostly of shrubs (from the Old World) with minute leaves. [Pg 69]

Genus **6. TÁMARIX.**

Leaves simple, very small, alternate, clasping; old ones almost transparent at the apex. Flowers in spike-like panicles, small, red, or pink, rarely white.

T. Gállica.

Támarix Gállica, L. (French Tamarisk.) Leaves very small, acute; spray very slender, abundant. A sub-evergreen shrub or small tree, 5 to 20 ft. high; with very small pinkish flowers, in spike-like clusters, blooming from May to October. A very beautiful and strange-looking plant, which, rather sheltered by other trees, can be successfully grown throughout.

Order **V. TERNSTRŒMIÀCEÆ.**

(Tea or Camellia Family.)

An order of showy-flowered trees and shrubs of tropical and sub-tropical regions, here represented by the following genera:

Genus **7. STUÁRTIA.**

Shrubs or low trees with alternate, simple, exstipulate, ovate, serrulate leaves, soft downy beneath. Flowers large (2 in.), white to cream-color, solitary and nearly sessile in the axils of the leaves; blooming in early summer. Fruit a 5-celled capsule with few seeds; ripe in autumn.

S. pentágyna.

1. **Stuártia pentágyna**, L'Her. (Stuartia.) Leaves thick, ovate, acuminate, acute at base, obscurely mucronate, serrate, finely pubescent, 3 to 4 in. long, one half as wide. Flowers whitish cream-colored, one petal much the smallest; stamens of the same color. Pod 5-angled. [Pg 70]

S. Virgínica.

Handsome shrub or small tree (10 to 15 ft.), wild south in the mountains, and hardy and cultivated as far north as New York City without protection. In Massachusetts it needs some sheltered position.

2. **Stuártia Virgínica**, Cav. (Virginia Stuartia.) Leaves elliptic-ovate, acuminate at both ends, 2 in. long, 1 in. wide, thin, serrate, silky pubescent beneath. Flowers white with purple filaments and blue anthers. Pod globular and blunt; ripe in October. A beautiful shrub rather than tree (8 to 12 ft.), wild in Virginia and south; hardy as far north as Washington.

Genus **8. GORDÒNIA.**

Shrubs or small trees with alternate, simple, feather-veined leaves. Flowers large (3 to 4 in. wide), white, showy, solitary in the axils of the leaves. Blooming in summer. Fruit a dry, dehiscent, conical-pointed, 5-celled capsule with 10 to 30 seeds, ripe in the autumn.

G. Lasiánthus.

1. **Gordònia Lasiánthus**, L. (Loblolly Bay.) Leaves thick, evergreen, lanceolate-oblong, minutely serrate, nearly sessile, smooth

and shining on both sides. The large, solitary, sweet-scented, axillary flowers on peduncles half as long as the leaves. A large tree (30 to 70 ft. high) in the south (wild in southern Virginia), and cultivated as far north as central Pennsylvania, without protection; at St. Louis and Boston it needs protection. Wood of a reddish color, light and brittle.

G. pubéscens.

2. **Gordònia pubéscens**, L'Her. Leaves thin, deciduous, obovate-oblong, sharply serrate, white beneath. Flowers nearly sessile. A small tree or shrub of the [Pg 71] south (30 ft. high in Georgia), hardy, and rarely cultivated as far north as Philadelphia, or still farther north if slightly sheltered.

Order **VI. MALVÀCEÆ.** (Mallow Family.)

A large family, mainly of herbs, found in tropical and temperate regions. One cultivated species, almost a tree, is included in this work.

Genus **9. HIBÍSCUS.**

Herbs or shrubs; one sometimes tree-like, with simple, deciduous, alternate, stipulate, usually lobed leaves. Flowers large, showy, 5-parted (Hollyhock-shaped), in late summer. Fruit a 5-celled, many-seeded pod, ripe in autumn.

H. Syrìacus.

Hibíscus Syrìacus, L. (Tree Hibiscus.) The only woody and sometimes tree-like species; has ovate, wedge-shaped, 3-lobed, toothed
leaves, and large (3 in.) white, purple, red, or variegated flowers.
Usually a shrub, 6 to 15 ft. high, often cultivated throughout; introduced from Syria.

Order **VII. STERCULIÀCEÆ.**

Trees or shrubs (a few are herbs), with alternate leaves, and the
stamens united into a tube. A large order of tropical plants.

Genus **10. STERCÙLIA.**

Leaves alternate, simple, usually lobed, ovaries more or less divided into 5 carpels, each 2- to many-lobed; fruit when ripe forming
a star of 5 distinct pods. [Pg 72]

S. platanifòlia.

Stercùlia platanifòlia, L. (Chinese Parasol.) Leaves large, deciduous, alternate, palmately 3- to 5-lobed, deeply heart-shaped at base,
the margin entire, the lobes acute; smooth or slightly hairy; leafstalk
about as long as the blade. Flowers green, in axillary panicles; fruit
star-shaped. A small, beautiful tree from China; probably not hardy
north of Washington.

Order **VIII. TILIÀCEÆ.** (Linden Family.)

An order, mainly of trees, abundant in the tropics; here represented by a single genus:

Genus **11. TÍLIA.**

Trees with alternate, deciduous, obliquely heart-shaped, serrate
leaves, about as broad as long. Leaves two-ranked on the stem.
Flowers small, cream-colored, fragrant, in clusters on a peculiar,
oblong, leaf-like bract. Fruit small (1/8 in.), globular, woody, in
clusters from the same bract. Wood white and soft; inner bark very
fibrous and tough.

* Flowers with petal-like scales among the stamens; American
species. (**A.**)

    A. Leaves very large, 6 to 8 in.   3.

    A. Leaves medium, 4 to 6 in.   1.

    A. Leaves small, 2 to 3 in.   2.

* Flowers with no petal-like scales among the stamens.   4.

T. Americàna.

1. **Tília Americàna**, L. (Basswood. Whitewood. Linden.) Leaves
large, 4 to 6 in. long, green and smooth, or very nearly so, thickish.
Fruit ovoid, somewhat ribbed, ¼ in. broad, greenish when ripe in
October, on a bract which is usually tapering to the base. Tall tree,
60 to 80 ft. high, wild in rich woods and often cultivated.

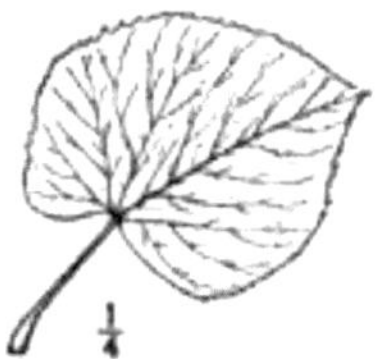

T. pubéscens.

2. **Tília pubéscens**, Ait. (Small-leaved Basswood.) Leaves smaller, 2 to 3 in. long, [Pg 73] thinner and rather pubescent beneath. Fruit globose, 1/5 in. broad, on a bract usually quite rounded at base.

This is usually considered as a variety of the last-named species. It is found from New York south and west.

T. heterophýlla.

3. **Tília heterophýlla**, Vent. (White Basswood.) Leaves large, often 8 in. broad, smooth and bright green above, silvery white and downy beneath, with darker, purplish veins. A large tree; wild in Pennsylvania, west and south, and often cultivated.

T. Europæa.

4. **Tília Europæa**, Mill. (European Linden.) Leaves twice as long as the petioles, and smooth except a woolly tuft in the axils of the veins beneath. Small and large leaved varieties are in cultivation.

The flowers have no petal-like scales among the stamens, while the American species have. An ornamental tree with dense foliage; often cultivated from Europe. The twigs are more numerous and more slender than those of the American species. Nearly a score of named varieties are in cultivation. Var. *laciniata* has deeply cut and twisted leaves.

Order **IX. RUTÀCEÆ.** (Rue Family.)

Shrubs and trees, rarely herbs, in most cases with transparent-dotted, heavy-scented foliage. A rather large order in warm climates.

Genus **12. XANTHÓXYLUM.**

Shrubs or trees with mostly odd-pinnate, alternate leaves. The stem and often the leaflets prickly; flowers small, greenish or whitish; fruit dry, thick pods, with 1 to 2 seeds.

X. Americànum.

1. **Xanthóxylum Americànum**, Mill. (Northern Prickly-Ash. Toothache-Tree.) Leaves and flowers in sessile, axillary, umbellate clusters; leaflets 5 to [Pg 74] 9, ovate-oblong, downy when young. Flowers appear before the leaves. Shrub, scarcely at all tree-like, with bark, leaves, and pods very pungent and aromatic. Common north, and sometimes cultivated.

X. Clàva Hércules.

2. **Xanthóxylum Clàva Hércules**, L. (Southern Prickly-Ash.) Leaflets 7 to 17, ovate to ovate-oblong, oblique at base, shining above. Flowers appear after the leaves. A small tree with very sharp prickles. Sandy coast of Virginia and southward; occasionally cultivated in the north.

Genus **13. PTÈLEA.**

Shrub with compound leaves of three leaflets, greenish-white flowers in terminal cymes, and 2-seeded fruit with a broad-winged margin, somewhat like the Elm, only larger.

P. trifoliàta.

Ptèlea trifoliàta, L. (Hop-Tree. Shrubby Trefoil.) Leaflets ovate, pointed, downy when young. Flowers with a disagreeable odor; fruit bitter, somewhat like hops. A tall shrub, often, when cultivated, trimmed into a tree-like form. Wild, in rocky places, in southern New York and southward.

Genus **14. PHELLODÉNDRON.**

Leaves opposite, odd-pinnate. Flowers diœcious; so only a portion of the trees bear the small, odoriferous, 5-seeded, drupe-like fruit.

P. Amurénse.

Phellodéndron Amurénse. (Chinese Cork-Tree.) Leaves opposite, odd-pinnate, 1½ to 3 ft. long; leaflets 9 to many, lanceolate, [Pg 75] sharply serrate, long-acuminate. Flowers inconspicuous, diœcious, in loose-spreading clusters at the ends of the branches. The pistillate flowers form small, black, pea-shaped fruit, in loose, grape-like clusters, thickly covered with glands containing a bitter, aromatic oil, and remaining on the tree in winter. Medium-sized tree (20 to 40 ft.), with Ailanthus-like leaves which turn bright red in autumn, and remain long on the tree. Hardy as far north as central Massachusetts.

Order **X. MELIÀCEÆ.** (Melia Family.)

Tropical trees, including the Mahogany; represented in the south by the following:

Genus **15. MÈLIA.**

Trees with alternate, bipinnate leaves. The flowers are conspicuous and beautiful, in large panicles, in the spring. Fruit in large clusters of berry-like drupes, with a 5-celled stone.

M. Azédarach.

Mèlia Azédarach, L. (China-Tree. Pride of India.) Leaves very large, doubly pinnate, with many obliquely lance-ovate, acuminate, smooth, serrate leaflets. Flowers small, lilac-colored, deliciously fragrant, in large axillary clusters. Fruit globular, as large as cherries, yellow when ripe in autumn; hanging on through the winter. A rather small (20 to 40 ft. high), rapidly growing, round-headed, popular shade-tree in the south, and hardy as far north as Virginia. Introduced from Persia. [Pg 76]

Genus **16. CEDRÉLA.**

Leaves large, alternate, deciduous, odd-pinnate. Flowers with separate petals, fragrant, white, in large clusters. Fruit 5-celled dehiscent pods, with many pendulous, winged seeds.

C. Sinénsis.

Cedréla Sinénsis. (Chinese Cedrela.) Leaves large, odd-pinnate, alternate, appearing much like those of the Ailanthus, but with slight serrations near the tips of the leaflets, and no glands near the base. Bruised leaves with a strong odor; footstalk and stout-tipped branches with glands. Large tree, seemingly hardy in New Jersey, but dies to the ground in winter in Massachusetts. Recently introduced from China.

Order **XI. SIMARUBÀCEÆ.** (Quassia Family.)

Eastern trees and shrubs, here represented by a single tree:

Genus **17. AILÁNTHUS.**

Large trees to shrubs, with alternate, odd-pinnate leaves. Flowers small, greenish, in large terminal panicles. Fruit broadly winged, like the Ash, but with the seed in the center.

A. glandulòsus.

Ailánthus glandulòsus, Desf. (Tree of Heaven.) Leaves very large, 2 to 5 ft. long on the younger growths; leaflets obliquely lanceolate, coarsely toothed at the base, with a gland on the lower side at the point of each tooth; point of leaflets entire. Young twigs thick, rusty brown; buds very small in the axils. Only some of the trees have fruit, as [Pg 77] some have only staminate flowers. The staminate flowers are very ill-scented. A rapid-growing tree, with useful hard wood; cultivated and naturalized; hardy throughout. See page 10.

Order **XII. ILICÌNEÆ.** (Holly Family.)

A small order of trees and shrubs, including for our purpose only one genus:

Genus **18. ÌLEX.**

Trees or shrubs with simple, alternate, thick, mostly evergreen leaves. Flowers rather inconspicuous, mostly in clusters. Fruit berry-like, small (¼ to ½ in.), with 4 to 6 nutlets; hanging on the plants late in the autumn or through the winter.

* Leaves evergreen. (**A.**)

    A. Leaves with spiny teeth   1.

    A. No spiny teeth   2.

* Leaves deciduous   3.

I. opàca.

1. **Ìlex opàca**, Ait. (American Holly.) Leaves evergreen, oval, acute, thick, smooth, with scattered spiny teeth. Flowers white; May. The bright-red berries, found only on some of the trees, remain on through the greater part of the winter. Small tree, 15 to 40 ft. high, with very hard white wood; wild in southern New England and southward. A beautiful broad-leaved, evergreen tree which should be more extensively cultivated. North of latitude 41° it needs a protected situation.

I. Dahòon.

2. **Ìlex Dahòon**, Walt. (Dahoon Holly.) Leaves 2 to 3 in. long, evergreen, oblanceolate or oblong, entire or sharply serrate toward the apex, with revolute margins, not spiny. Young branches and [Pg 78] lower surface of the leaves, especially on the midrib, pubescent. Small tree, 10 to 30 ft. high; Virginia and south, with very hard, white, close-grained wood. Rarely cultivated.

I. montícola.

3. **Ìlex montícola**, Gray. Leaves deciduous, ovate to lance-oblong, 3 to 5 in. long, taper-pointed, thin, smooth, sharply serrate. Fruit red, on short stems, with the seeds many-ribbed on the back. Usually a shrub but sometimes tree-like; damp woods in the Catskills and in the Alleghany Mountains.

Order **XIII. CELASTRÀCEÆ.**

Shrubs with simple leaves and small, regular flowers, forming a fruit with ariled seeds.

Genus **19. EUÓNYMUS.**

Shrubs somewhat tree-like, with 4-sided branchlets, opposite, serrate leaves, and loose cymes of angular fruit which bursts open in the autumn.

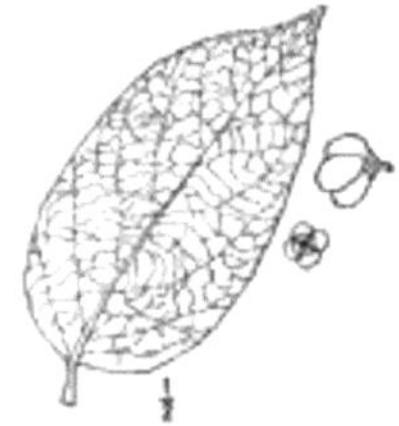

E. atropurpùreus.

1. **Euónymus atropurpùreus**, Jacq. (Burning-bush. Wahoo.) Leaves petioled, oval-oblong, pointed; parts of the dark-purple flowers commonly in fours; pods smooth, deeply lobed, when ripe, cinnamon in color and very ornamental. Tall shrub, 6 to 20 ft. high; wild in Wisconsin to New York, and southward; often cultivated.

E. Europæus.

2. **Euónymus Europæus**, L. (European Spindle-tree or Burning-bush.) Leaves oblong-lanceolate, serrate, smooth; flowers and fruit commonly in threes on compressed stems; fruit usually 4-lobed, the lobes acute; flowers greenish-white; May; fruit abundant, scarlet, ripe in September. [Pg 79] Generally a shrub, though sometimes tall enough (4 to 20 ft.) and trimmed so as to appear tree-like; twigs smooth, green or reddish-green. Extensively cultivated; from Europe.

Order **XIV. RHAMNÀCEÆ.**

(Buckthorn Family.)

An order mainly of shrubs, but including in the north-eastern United States two or three small trees.

Genus **20. RHÁMNUS.**

Shrubs or small trees with deciduous (rarely evergreen), usually alternate (rarely opposite), pinnately veined leaves. Flowers small, 4-parted, inconspicuous, in clusters in the axils of the leaves. Fruit berry-like, with 2 to 4 seed-like nuts.

* Branches terminating in thorns   1.

* Plant without thorns. (**A.**)

    A. Leaves deciduous   2.

    A. Leaves evergreen   3.

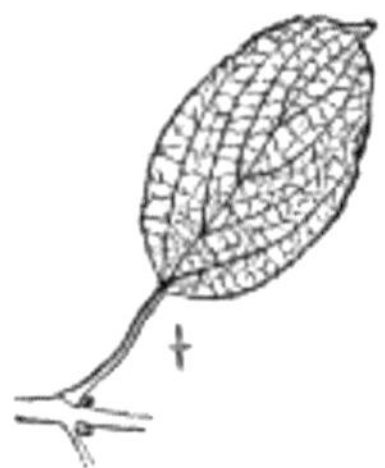

R. cathártica.

1. **Rhámnus cathártica, L.** (Common Buckthorn.) Leaves ovate, minutely serrate, alternate or many of them opposite; branchlets terminating in thorns. Flowers greenish. Fruit globular, 1/3 in. in diameter, black with a green juice, and 3 or 4 seeds; ripe in September. A shrub or small tree, 10 to 15 ft. high, from Europe; cultivated for hedges, and found wild in a few places, where it forms a small tree.

R. Caroliniàna.

2. Rhámnus Caroliniàna, Walt. (Carolina Buckthorn.) Leaves 3 to 5 in. [Pg 80] long, alternate, oblong, wavy and obscurely serrulate, nearly smooth, on slender pubescent petioles. Flowers greenish, 5-parted, solitary or in umbellate clusters in the axils. Fruit berry-like, globular, the size of peas, 3-seeded, black when ripe in September. A thornless shrub or small tree, 5 to 20 ft. high. New Jersey, south and west. Usually a shrub except in the Southern States.

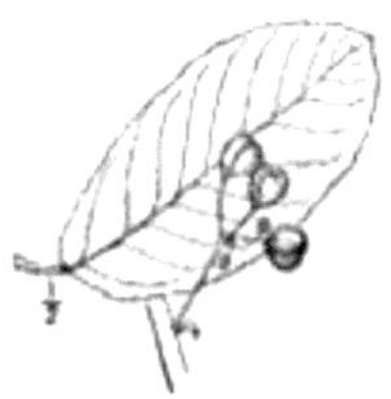

R. Califòrnicus.

3. **Rhámnus Califòrnicus**, Esch. (California Buckthorn.) Leaves evergreen, oval-oblong to elliptical, 1 to 4 in. long, rather obtuse, sometimes acute, generally rounded at base, serrulate or entire. Fruit blackish purple, with thin pulp, ¼ in., 2- to 3-seeded. A spreading shrub, 5 to 18 ft. high, without thorns; from California.

Genus **21. HOVÈNIA.**

Leaves alternate, deciduous, simple, oblique at base. Fruit an obscurely 3-lobed, 3-celled, 3-seeded pod in dichotomous clusters, both axillary and terminal.

H. dúlcis

Hovènia dúlcis, Thunb. Leaves long-petioled, more or less ovate to cordate, serrate, palmately 3-ribbed, much darker on the upper surface; both sides slightly roughened with scattered hairs. Fruit sweet, edible, in clusters in the axils of the leaves; seeds lens-shaped, with a ridge on the inner side. Flowers white; in July. A large, broad-topped tree, introduced from Japan. Hardy at Washington, but dies to the ground in the Arnold Arboretum, Massachusetts.

Genus **22. ZÌZYPHUS.**

Leaves simple, alternate, deciduous, 3-ribbed. Flowers axillary, 5-petaled. Fruit fleshy, drupe-like, containing a 1- to 2-celled nut.

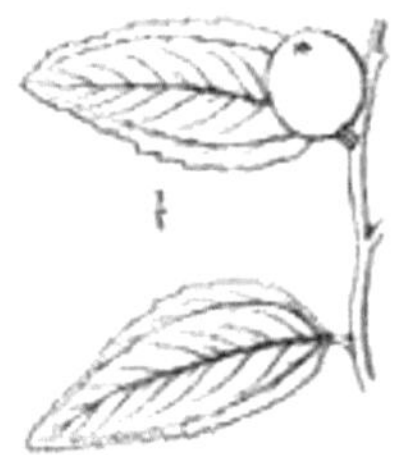

Z. vulgàris.

Zìzyphus vulgàris, Lam. (Jujube.) Leaves ovate-lanceolate, obtuse, serrate, smooth, and glossy green on both sides, upper side [Pg 81] quite dark; slightly hairy beneath on the veins; prickles twin, one recurved, sometimes none. New growth of the year green, and resembling a once-pinnate compound leaf and usually dropping off in the autumn like one. Leaves 10 to 20 on a twig, 2-ranked; flowers and drupes nearly sessile in the axils; fruit small (¼ in.), blood-red when ripe. A small tree (10 to 30 ft. high), of recent introduction from Syria; hardy at Philadelphia, but needing some protection at the Arnold Arboretum, Massachusetts.

Order **XV. SAPINDÀCEÆ.** (Soapberry Family.)

A large order represented in all countries, and so varied in its characteristics as to form several sub-orders.

Genus **23. ÆSCULUS.**

Deciduous trees or sometimes shrubs, with opposite, palmately compound leaves with serrated, straight-veined leaflets. Flowers usually conspicuous in dense terminal panicles. Fruit large, leathery-coated, often rough, with one or few large Chestnut-like but bitter seeds. Fruit large in midsummer, hanging on the tree until frost.

* Fruit prickly. (**A.**)

A. Leaflets usually 7; flowers widely spreading		1.
A. Leaflets 5-7, red-spotted and rough; flowers rosy red	*Æsculus rubicunda*	(1).
A. Leaflets usually 5; flowers not much sprea-		2.

ding

* Fruit smooth or nearly so. (**B.**)

  B. Flowers bright red   3.

  B. Flowers yellow, purplish or pinkish   4.

  B. Flowers white, in long, slender, erect clusters   5.

Æ. Hippocástanum.

1. **Æsculus Hippocástanum.** (Common Horse-chestnut.) Leaves of 7 obovate, abruptly pointed, serrated leaflets. Flowers [Pg 82] very showy in large clusters, with 5 white, purple and yellow spotted, broadly spreading petals. A variety with double flowers is in cultivation. May or June. Fruit large, covered with prickles. Seeds large, chestnut-colored. Tree of large size, with brown twigs; cultivated everywhere; from Asia.

Æ. rubicúnda.

Æsculus rubicunda (Red-flowering Horse-chestnut) is frequent in cultivation; leaflets 5 to 7, red-spotted and rough; flowers rosy red. It is probably a hybrid between the common Horse-chestnut and one of the Buckeyes.

Æ glàbra.

2. **Æsculus glàbra**, Willd. (Ohio Buckeye.) Leaves with 5 oval-oblong, acuminate, serrate, smooth leaflets. Flowers not showy, yellowish-white, with 4 somewhat irregular, slightly spreading petals. June. Fruit small, 1 in. in diameter, covered with prickles, at least when young; ripe in autumn. Small to large tree, wild in the basin of the Ohio River, along river-banks. Sometimes cultivated.

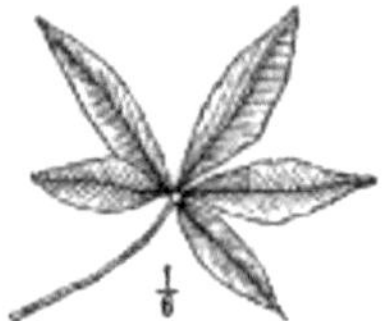

Æ. Pàvia.

3. **Æsculus Pàvia**, L. (Red Buckeye.) Leaves of 5 to 7 oblong-lanceolate, finely serrate, generally smooth leaflets, of a shining green color, with purple veins and petioles. Flowers (corolla and calyx) bright red, with included stamens; corolla of 4 petals, not spreading; calyx tubular. Fruit smooth, oblong-obovate, 1 in. long. Small tree or shrub, 10 to 20 ft. high, with purple twigs. Virginia west and south, and occasionally cultivated throughout.

Æ. flàva.

104

4. **Æsculus flàva**, Ait. (Sweet Buckeye.) Leaves with 5 to 7 serrulate, elliptical, acuminate leaflets, usually smooth, sometimes [Pg 83] minutely pubescent beneath; the pubescent petiole flattish toward the base. Flowers yellow, not spreading. Spring. Fruit globose, uneven but not prickly, 2 in. in diameter. Seeds large (1 in.), 1 or 2 in number, mahogany-colored; ripe in autumn. Often a large tree, sometimes only a shrub, 6 to 70 ft. high, in rich woods; Virginia to Indiana, and southward. Cultivated occasionally throughout.

Var. *purpurascens* of this species has flesh-colored or dull-purple flowers, and leaflets quite downy beneath.

Æ. macrostàchya.

5. Æsculus macrostàchya, Mx. (Long-racemed Buckeye.) Leaflets 5 to 7, ovate, acuminate, serrate, velvety with hairs beneath. Flowers white, in long, slender, erect clusters; July; petals 4, spreading; stamens very long. A beautiful, widely spreading shrub. 5 to 18 ft. high; from the Southern States; often cultivated. Probably hardy throughout.

Genus 24. KŒLREUTÈRIA.

A small tree with alternate, once to twice irregularly pinnate leaves with many coarsely toothed leaflets. Flowers conspicuous, yellow, in terminal panicles. In summer. Fruit rounded, bladdery, 3-celled, few-seeded pods; ripe in autumn.

K. paniculàta.

Kœlreutèria paniculàta, Laxm. Leaflets thin and very irregularly toothed. Clusters 6 to 12 in. long, of many irregular flowers, ½ in. wide; through the summer. Fruit an ovate, bladdery capsule, ripening in autumn. A fine, small, round-headed tree, 20 to 40 ft. high; from China. Probably hardy throughout. [Pg 84]

Genus **25. ÀCER.**

Trees, or rarely shrubs, with simple, opposite, and almost always palmately lobed leaves, which, in our species, are always deciduous. Flowers small and usually dull-colored, in clusters. Fruit double-winged and 2-seeded, in some species hanging on the tree till the leaves have fallen; in others dropping off early in the spring. The species differ much in the spreading of the wings of the fruit. Wood light-colored and medium hard; bark rather smoothish, but in large trees with longitudinal cracks.

* Leaves slightly or not lobed   13.

* Leaves about 3-lobed (rarely 5-lobed); shrubs or small trees. (**A.**)

  A. Leaves serrate   1, 2.

  A. Leaves somewhat sinuate, not at all serrate; juice milky.   10.

* Leaves 5-, rarely 3-lobed. (**B.**)

  B. The lobes acute, irregularly but quite fully serrate; juice not milky. (**C.**)

    C. The fruit in corymbs, dropping early; American species. (**D.**)

      D. Leaf-notches somewhat rounded; tree large; limbs drooping on old trees   3.

D. Leaf-notches acute; tree small   4.

C. Fruit in hanging racemes, remaining on the tree till autumn; leaves thickish   5.

B. The lobes acute; sparingly or not at all serrate. (**E.**)

E. Juice not milky   6.

E. Juice milky at the bases of the leaves   8, 9.

B. The lobes obtuse and sinuate   10.

* Leaves 5- to 7-lobed. (**F.**)

F. Lobes fully serrate   11.

F. Lobes sparingly serrate. (**G.**)

G. Juice milky   8, 9.

G. Juice not milky; leaves 8 to 10 in. broad   7.

F. Lobes somewhat sinuate, not serrate; juice milky   10.

* Leaves with 7 or more lobes   11, 12.

À. spicàtum.

1. **Àcer spicàtum**, Lam. (Mountain Maple.) Leaves with 3 (rarely 5) coarsely serrated, taper-pointed lobes, with slightly cordate base; downy beneath. Flowers greenish-yellow, in erect, [Pg 85] slender racemes or panicles, blooming in June. Wings of the small fruit at about a right angle. Small tree, 6 to 10 ft. high, or usually a shrub, with brown twigs. Native; growing in moist woods; rarely cultivated.

À. Pennsylvánicum.

2. **Àcer Pennsylvánicum**, L. (Striped Maple.) Leaves large, thin, 3-lobed at the end, cordate at base, finely and sharply doubly serrate. Flowers greenish, in drooping, elongated, loose racemes appearing after the leaves in spring. Fruit with large diverging wings. A small, slender tree, with light green bark striped with dark red. Wild throughout and cultivated.

À. dasycárpum.

3. **Àcer dasycárpum**, Ehrh. (Silver or White Maple.) Leaves large, truncated at base, 5-lobed, with blunt notches, the lobes irregularly serrated and notched, silvery white, and, when young, downy beneath. Flowers light yellowish-purple, preceding the leaves, in crowded umbels along the branches. Wings of fruit large and forming about a right angle; ripe early in June. A rather large, rapidly growing, and usually somewhat weeping tree, with soft white wood. Special cut-leaved and weeping varieties are sold at the nurseries. Wild along river-banks, and extensively cultivated in the streets of cities.

À. rùbrum.

4. **Àcer rùbrum**, L. (Red Maple.) Leaves cordate at base and cleft into 3 to 5 acute-notched, irregularly toothed lobes, whitish beneath, turning a bright crimson in early autumn. Flowers usually scarlet, rarely yellowish, in close clusters along the branches, appearing before the leaves in the spring. [Pg 86] Fruit often reddish, small, with the wings at about a right angle. A rather small, somewhat spreading tree with reddish branches; wild in wet places and often cultivated.

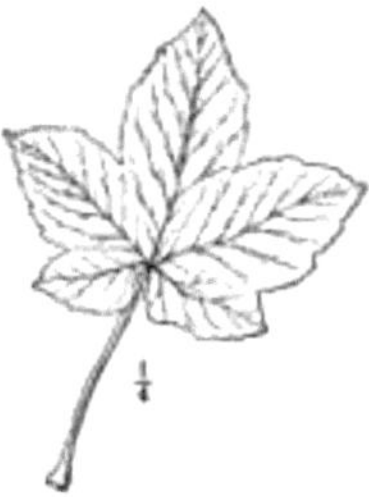

À. Pseudoplátanus.

5. **Àcer Pseudoplátanus**, L. (Sycamore-maple.) Leaves thickish, cordate, downy beneath, with 5 rather crenately toothed lobes, on long, often reddish petioles. Flowers in long pendulous racemes, appearing after the leaves. Fruit hanging on the tree till after the leaves fall in the autumn, the wings forming about a right angle. A rather large, spreading tree, 30 to 80 ft. high, with reddish-brown twigs. Cultivated; from Europe. Many varieties of this species are sold by the nurserymen; among them may be mentioned the Purple-leaved, Golden-leaved, Silver-leaved, Tricolored, etc.

À. saccharìnum.

6. **Àcer saccharìnum**, Wang. (Sugar or Rock Maple.) Leaves deeply 3- to 5-lobed, with rounded notches; lobes acute, few-toothed; base heart-shaped, smooth above, glaucous beneath. Flowers hanging in umbel-like clusters at the time the leaves are expanding in the spring. Fruit with wings not quite forming a right angle. A large (50 to 100 ft. high), very symmetrical tree, ovate in form, with whitish-brown twigs. Wild throughout, and extensively cultivated in the streets of cities.

Var. *nigrum*, Torr. and Gray. (Black Sugar-maple.) Leaves scarcely paler beneath, but often minutely downy; lobes wider, often shorter and entire; notch at the base often closed (the under leaf in the figure). Found with the other Sugar-maple, and quite variable.

À. macrophýllum.

7. **Àcer macrophýllum**, Ph. (Large-leaved or California Maple.) Leaves very large, 8 to 10 in. broad; 5-, sometimes 7-lobed, with deep, rounded notches; lobes themselves somewhat 3-lobed and repand-notched; pubescent beneath. Flowers yellow, in erect panicles, fragrant, blooming after the leaves are expanded. Fruit large, with [Pg 87] the seeded portion hairy; wings at about a right angle. Tree very large (100 ft. high); wood soft, whitish, beautifully veined.

Twigs brown; buds green. Cultivated; from the Pacific coast, but not hardy north of 40° N. latitude.

À. platanoìdes.

8. **Àcer platanoìdes**, L. (Norway Maple.) Leaves large, smooth, 5-, rarely 7-cleft, with cordate base; lobes acute, with few coarse, sharp teeth, bright green both sides. The leaves resemble those of the Sycamore (Platanus). Flowers a little later than the leaves in spring, in stalked corymbs, less drooping than the Sugar-maple (No. 6). Fruit with wings diverging in a straight line. A medium-sized, broad, rounded tree with brown twigs and milky juice, best seen at the bases of the young leaves. Cultivated throughout.

À. Læ̀tum.

9. **Àcer Læ̀tum.** (Colchicum-leaved Maple.) Leaves 5- to 7-lobed, scarcely heart-shaped at base, smooth and green on both sides; juice milky; the lobes usually without any notches or irregularities, sometimes with about three winding sinuations. Flowers in erect corymbs. Differs from Acer platanoides in having the lobes of the leaves more nearly entire, and the fruit much smaller with wings not so broadly spreading.

À. campéstre.

10. **Àcer campéstre**, L. (English or Cork-bark Maple.) Leaves cordate, with usually 5 roundish lobes, sparingly crenate or rather undulated; juice milky. Racemes of flowers erect, appearing after the leaves in spring. Wings of the fruit [Pg 88] broadly spreading; fruit ripening very late. A low (15 to 30 ft. high), round-headed tree, with the twigs and smaller branches covered with corky bark. Occasionally cultivated; from Europe.

Var. *variegatum* has white blotched leaves.

À. palmàtum.

11. **Àcer palmàtum**, Thunb. (Palmate-leaved Japan Maple.) Leaves small, smooth, palmately parted into 5 to 9 quite regularly serrated lobes. Flowers in small umbels. A very low tree, almost a shrub; cultivated; from Japan; probably hardy throughout. There are a great number of Japan Maples, many of them probably varieties of this species, others hybrids. The leaves of some are so divided and dissected as to form merely a fringe or feather. In color they range from pure green to the richest reds.

112

À. circinàtum.

12. **Àcer circinàtum**, Pursh. (Round-leaved or Vine Maple.) Leaves orbicular, with 7 to 11 serrated, acute lobes, a heart-shaped base, reddish-green color, and both surfaces smooth. Corymbs of purplish flowers, small and hanging on long peduncles; appearing after the leaves. Wings of the fruit diverging in a straight line. A small tree or tall shrub, 10 to 30 ft. high, of spreading habit, with smooth bark, and pale brown twigs; cultivated; from the Pacific coast of North America.

À. Tartáricum.

13. **Àcer Tartáricum**, L. (Tartarian Maple.) Leaves ovate, slightly cordate, rarely lobed, serrated, light-colored, expanding very early in the spring. Panicle of greenish-yellow flowers erect, blooming after the leaves have expanded. Wings of the fruit parallel or sometimes touching. A small tree, sometimes shrubby in growth, of irregular form, with brown twigs; rarely cultivated; from Europe.

Genus **26. NEGÚNDO.**

Leaves pinnate, of 3 to 5 leaflets. Flowers rather inconspicuous. Fruit a two-winged key as in Acer, in drooping racemes. [Pg 89]

N. aceroìdes.

Negúndo aceroìdes, Moench. (Ash-leaved Maple. Box-elder.) Leaves pinnate, of 3 to 5 (rarely 7) coarsely and sparingly toothed leaflets. Flowers staminate and pistillate on separate trees, in drooping clusters rather earlier than the leaves. Fruit on only a portion of the trees; wings forming less than a right angle. A rather small (30 to 60 ft. high), rapidly growing tree, with light pea-green twigs; wild from Pennsylvania and south, and cultivated throughout.

Var. *Californicum*, Torr. and Gray (the under drawing in the figure), has leaflets more deeply cut, thicker, and quite hairy; it is occasionally cultivated.

Order **XVI. ANACARDIÁCEÆ.**

(Cashew Family.)

Trees and shrubs, mainly of the tropical regions, here represented by only one genus:

Genus **27. RHÚS.**

Low trees or shrubs with acrid, often poisonous, usually milky juice, and dotless, alternate, usually pinnately compound leaves. Flowers greenish-white or yellowish, in large terminal panicles. Fruit small (1/8 in.), indehiscent, dry drupes in large clusters, generally remaining on through the autumn.

* Leaves simple, rounded, entire   6, 7.

* Leaves once-pinnate. (**A.**)

    A. Twigs very hairy; rachis not winged; leaflets 11 to 31   1.

    A. Twigs downy; rachis wing-margined; leaflets entire or nearly so   3.

A. Twigs smooth. (**B.**)

  B. Rachis of leaf broadly winged; leaflets serrate   5.

  B. Rachis not winged. (**C.**)

    C. Leaflets 11 to 31, serrate; fruit hairy   2.

    C. Leaflets 7 to 13, entire; fruit smooth; poisonous   4.

* Leaves twice-pinnate; variety under   2.

[Pg 90]

R. týphina.

1. **Rhús týphina**, L. (Stag-horn Sumac.) Leaflets 11 to 31, oblong-lanceolate, pointed, serrate (rarely laciniate), pale beneath. Branches and footstalks densely hairy. Fruit globular, in large, dense, erect panicles, covered with crimson hairs. Shrub or tree, 10 to 30 ft. high. It is very common along fences and on hillsides. The wood is orange-colored and brittle.

R. glàbra.

2. **Rhús glàbra**, L. (Smooth Sumac.) Leaflets 11 to 31, lanceolate-oblong, pointed, serrate, smooth, glaucous white beneath. Branches not hairy. Fruit globular, in a rather open, spreading cluster, covered densely with crimson hairs. A shrubby plant, 2 to 12 ft. high, found quite abundantly in rocky or barren soil throughout.

R. laciniàta.

Var. *laciniata* is frequently planted for ornament. It has very ir-regularly twice-pinnate leaves drooping gracefully from the branches.

R. copallìna.

3. **Rhús copallìna**, L. (Dwarf Mountain Sumac.) Branches and stalks downy; leafstalk wing-margined between the 9 to 21 oblong-lanceolate, usually entire leaflets, which are oblique at base and smooth and shining above. Wild in rocky hills throughout; often cultivated. North, a beautiful shrub; south, a tree. 2 to 25 ft. high.

R. venenàta.

4. **Rhús venenàta**, DC. (Poison-sumac. Poison-dogwood. Poison-elder.) Leaflets 7 to 13, obovate-oblong, entire, abruptly pointed, smooth or nearly so. Fruit small, globular, smooth, dun-colored, in loose axillary [Pg 91] panicles hanging on late in winter; the stone striate. This is a very poisonous species (to the touch), 6 to 18 ft. high, growing in swamps. Rarely at all tree-like.

R. Osbéckii.

5. **Rhús Osbéckii**, DC. (Chinese Sumac.) Leaves very large, pinnate, assuming in autumn a rich reddish-fawn or orange color; the leafstalk broadly winged between the leaflets; leaflets serrate. A small ornamental tree, 10 to 25 ft. high; cultivated; from China; quite hardy in the Northern States.

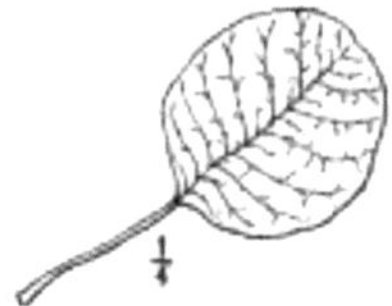

R. Cótinus.

6. **Rhús Cótinus**, L. (Smoke-tree. Venetian Sumac.) Leaves smooth, obovate, entire, on slender petioles. Flowers greenish, minute, in terminal or axillary panicles. Fruit seldom found. Usually most of the flowers are abortive, while their pedicels lengthen, branch, and form long feather-like hairs, making large cloud-like branches that look somewhat like smoke (whence the name). A shrub or small tree, 6 to 10 ft. high, often planted for ornament; from Europe.

R. cotinoìdes.

7. **Rhús cotinoìdes**, Nutt. (American Smoke-tree.) Leaves thin, oval, obtuse, entire, acute at base, 3 to 6 in. long, smooth or nearly so. Flowers and fruit like those of the cultivated species (Rhus Cotinus). A tree 20 to 40 ft. high; stem sometimes a foot or more in diameter in the Southern States; wild in Tennessee, west and south. Rare in cultivation. [Pg 92]

Order **XVII. LEGUMINOSÆ.** (Pulse Family.)

A very large order of plants, mainly herbaceous; found in all climates. A few are shrubby, and others are from small to large trees.

Genus **28. LABÚRNUM.**

Low trees or shrubs with alternate, palmate leaves of three leaflets. Flowers conspicuous, pea-blossom-shaped, in long hanging racemes, in late spring. Fruit pea-pod-shaped, dark brown, and many-seeded; ripe in autumn.

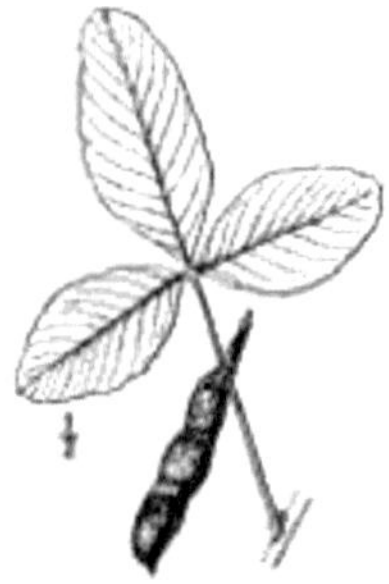

L. vulgàre.

Labúrnum vulgàre. (Laburnum. Golden-chain. Bean-trefoil Tree.) Leaves petiolate, with 3 ovate-lanceolate leaflets, pubescent beneath.

118

Flowers bright yellow, nearly 1 in. long, in long (1 ft.), pendulous, simple racemes; in late spring. Pods 2 in. long, linear, many-seeded, covered with closely appressed pubescence; one edge thick; ripe in autumn. A low, very ornamental tree, 10 to 20 ft. high, often cultivated; from Switzerland. Varieties with reddish, purple, and white flowers are also in cultivation.

Var. *alpinus* has smooth pods.

Genus 29. CARAGÀNA.

Leaves alternate, deciduous, abruptly once-pinnate; leaflets mucronate; stipules usually spinescent. Flowers pea-flower-shaped, mostly yellow. Trees or shrubs of Asia.

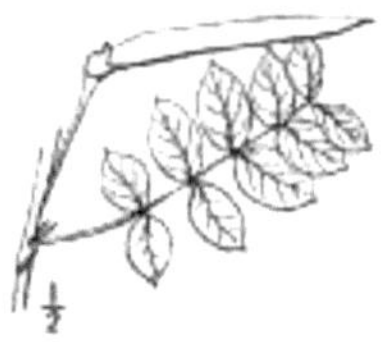

C. arboréscens.

Caragàna arboréscens, Larn. (Pea-tree.) Leaves with 4 to 6 pairs of oval-oblong, mucronate-pointed, hairy leaflets; petioles unarmed; stipules spinescent. Flowers yellow, blooming in May. Pods brown, ripe in August. A low, stiff, erect tree, 10 to 15 ft. high; in poor soil a bush. From Siberia; frequent in cultivation. [Pg 93]

Genus 30. CLADRÁSTIS.

Small tree with alternate, odd-pinnate leaves, the base of the petiole hollow, and inclosing the leaf-buds of the next year. Flowers large, pea-blossom-like in shape, in large clusters. Fruit pea-pod-like in shape and size. Wood light yellow, firm and hard.

C. tinctòria.

Cladrástis tinctòria, Raf. (Yellow-wood.) Leaflets 7 to 11, oval to ovate, 3 to 4 in. long, beautiful light green in color. Flowers 1 in. long, white, not so fragrant as the common Locust, in hanging panicles 10 to 20 in. long; blooming in June. Pods 2 in. long, ripe in August. Wild but rare in Kentucky and south. A beautiful tree, 20 to 50 ft. high, with very smooth grayish bark; rarely cultivated.

Genus **31. ROBÍNIA.**

Trees or shrubs with alternate, odd-pinnate leaves, having spines on each side of the stalk in place of stipules. Leafstalk thickened near the base, and covering 2 to 3 buds for the growth of a branch for the next year. An axillary bud also found that may produce a branch the same year as the leaf. Flowers large, pea-blossom-shaped, in large clusters. Fruit a pea-shaped pod.

* Branchlets and leafstalks not sticky   1.

* Branchlets and leafstalks sticky   2.

R. Pseudacácia.

1. **Robínia Pseudacácia**, L. (Common Locust.) Leaflets 9 to 19, small, oblong-ovate, entire, thin. Twigs purplish-brown, slender, smooth, not sticky. Flowers white, fragrant, in hanging racemes, 3 to 6 in. long. June. Pods flat, smooth, purplish-brown, ripe in September. An irregu [Pg 94] larly growing, slender tree, 70 to 80 ft. high, with white or greenish-yellow, very durable wood, and on old trees very rough bark with long, deep furrows. Native; Pennsylvania, west and south, and extensively planted and naturalized throughout. A number of varieties, some of which are thornless, are in cultivation.

R. viscòsa.

2. **Robínia viscòsa**, Vent. (Clammy Locust.) Leaflets 11 to 25, ovate-oblong, sometimes slightly heart-shaped at base, tipped with a short bristle. Twigs and leafstalks sticky to the touch. Flowers in a short, rather compact, upright raceme, rose-colored and inodorous. A small tree, 30 to 40 ft. high; native south, and has been quite extensively cultivated north.

3. **Robínia híspida**, L. (Bristly Locust. Rose-acacia.), with bristly leafstalks and branchlets, and large rose-colored flowers, is only a bush. Often cultivated. Wild from Virginia and south.

Genus **32. CÉRCIS.**

Small trees or shrubs, with alternate, simple, heart-shaped leaves. Flowers in umbel-like clusters along the branches, appearing before the leaves, and shaped like pea-blossoms. Fruit pea-like pods, remaining on the tree throughout the year. Wood hard, heavy, and beautifully blotched or waved with black, green, and yellow, on a gray ground.

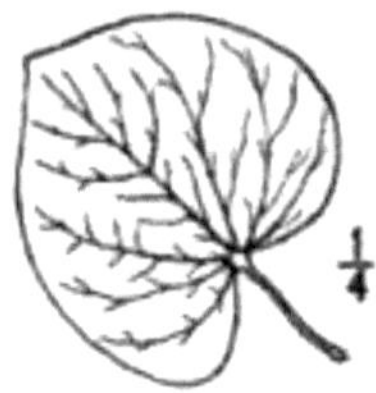

C. Canadénsis.

1. **Cércis Canadénsis**, L. (Judas-tree. Redbud.) Leaves acutely pointed, smooth, dark green, glossy. Flowers bright red-purple. Pods nearly sessile, 3 to 4 in. long, brown when ripe in August. A small ornamental tree, 10 to 30 ft. high, with smooth bark and hard apple-tree-like wood; wild from Central New York southward, and often cultivated.

2. **Cércis siliquástrum** (European Judas-tree.), from Europe, with obtusely pointed, somewhat kidney-shaped leaves, and white to purple flowers, is sometimes cultivated. It is not so tall or tree-like as the American species. [Pg 95]

Genus **33. GYMNÓCLADUS.**

Tall trees with alternate, very large (2 to 4 ft. long), unequally twice-pinnate leaves. Flowers white, conspicuous, in racemes at the ends of the branches. Fruit a large pea-like pod. Some trees are without fruit through the abortion of the pistils.

G. Canadénsis.

Gymnócladus Canadénsis, Lam. (Kentucky Coffee-tree.) Leaves 2 to 3 ft. long, often with the lower pinnæ simple and the upper pinnate. Leaflets ovate, of a dull bluish-green color. Shoots cane-like, blunt and stubby, quite erect. Bark exceedingly rough. Pod large, 6 to 10 in. long, 2 in. broad, with seeds over ½ in. across. A large (50 to 80 ft. high) tree with compact, tough, reddish wood. Wild from western New York southwestward, and occasionally cultivated as an ornamental tree.

Genus **34. GLEDÍTSCHIA.**

Usually thorny trees with alternate, once to twice abruptly pinnate leaves. Flowers inconspicuous, greenish, in small spikes. Summer. Fruit a small or large pea-like pod, with one to many seeds; ripe in autumn, but often hanging on the trees through the winter.

G. triacánthos.

1. **Gledítschia triacánthos**, L. (Honey-locust.) Leaflets lanceolate-oblong, somewhat serrate. Pods linear, 1 to 1½ ft. long, often twisted, filled with sweet pulp between the seeds. A large, handsome, clean tree, with usually many stout, much-branched thorns, especially abundant on bruised portions of the trunk and large branches; thorns compressed at base. Wild from Pennsylvania southward and westward, and extensively cultivated throughout.

A variety without thorns is frequently met with (var. *inermis*), also one with drooping foliage (var. *Bujotii pendula*). [Pg 96]

G. aquática.

2. **Gledítschia aquática**, Marsh. (Water-locust.) Leaflets ovate or oblong. Pods oval, 1 to 4 in. long, 1- to few-seeded, without pulp. A small tree with few slender, usually simple thorns; in swamps in southern Illinois and south. Occasionally planted for ornament. This species is quite similar to the preceding one, but the leaves are somewhat smaller, the thorns, though occasionally branching, do not branch so extensively, and the pod is very short and rounded.

G. sinénsis.

3. **Gledítschia sinénsis**, Lam. (Chinese Honey-locust.) A tree with stouter and more conical thorns, broader and more oval leaflets. A medium-sized or small tree, often cultivated. This species, like the others, has a thornless variety.

Genus **35. ALBÍZZIA.**

Trees or shrubs with abruptly pinnate leaves. Fruit a broad-linear straight pod.

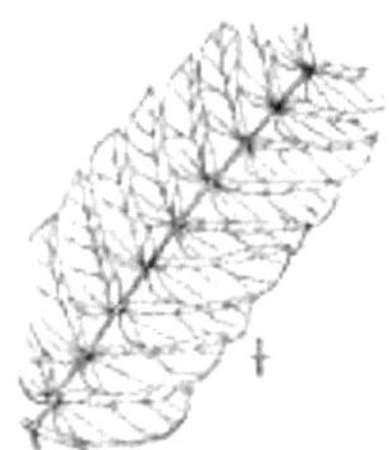

A. julibríssin.

Albízzia julibríssin, Boivin. (Silk-tree.) Leaves twice abruptly pinnate, of many (over 400) leaflets; leaflets semi-oblong, curved, entire, acute, with the midrib near the upper edge. Flowers in globose heads forming panicles. Fruit plain pods on short stems. A very beautiful small tree, introduced from Japan; probably not hardy north of Washington. The figure shows only one of the lowest and shortest side divisions (pinnæ) of the leaf. The pinnæ increase in length and number of leaflets to the end of the leaf. [Pg 97]

Order **XVIII. ROSACEÆ.** (Rose Family.)

A large and very useful order of trees, shrubs, and herbs of temperate regions.

Genus **36. PRÙNUS.**

Trees or shrubs with simple, alternate, deciduous, usually serrate, stipulate leaves, without lobes. The stems produce gum when injured. Foliage and nuts have flavor of peach-leaves. Flowers conspicuous, usually white, or light pink, often in clusters, peach-blossom-shaped; in early spring. Fruit in size from pea to peach, a rounded drupe with one stony-coated seed.

* Drupe large, soft velvety on the surface; stone rough (Peach, Apricot) — 1.

* Drupe medium, covered with a bloom; stone smooth, flattened (Plums). (**A.**)

   A. Usually thorny; wild, rarely cultivated. (**B.**)

      B. Leaves acuminate — 2, 3.

      B. Leaves not acuminate — 4, 5.

   A. Not thorny; cultivated — 6.

* Drupe medium to small, smooth, without bloom (Cherries). (**C.**)

   C. Drupes clustered in umbels, ½-1 in. in diameter. (**D.**)

      D. Small cultivated tree; drupe globose, rather large, very sour — 9.

      D. Large cultivated tree; drupe large, somewhat pitted at the stem — 8.

      D. Rather small, native tree; drupe small, flesh thin — 7.

   C. Drupes clustered in racemes, 1/8 - 1/3 in. in diameter. (**E.**)

      E. Tall shrubs rather than trees; racemes short — 11.

      E. Trees; racemes quite elongated. (**F.**)

F. Stone of fruit somewhat roughened   12.

F. Stone smooth   10.

P. Pérsica.

1. **Prùnus Pérsica**, L. (Common Peach.) Leaves lanceolate, serrate. Flowers rose-colored, nearly sessile, very early in bloom. Fruit clothed with velvety down, large; stone rough-wrinkled. A small tree, 15 to 30 ft. high, cultivated in numberless varieties for its fruit. Var. *lævis* (Nectarine) has smooth-skinned fruit. [Pg 98]

P. Americàna.

2. **Prùnus Americàna**, Marsh. (Wild Yellow or Red Plum.) Leaves ovate or somewhat obovate, conspicuously pointed, coarsely or doubly serrate, very veiny, smooth when mature. Fruit with little or no bloom, ½ to 1 in. in diameter, yellow, orange, or red; skin tough and bitter. Stone with two sharp edges. A small, thorny tree, 8 to 20 ft. high, common in woodlands and on river-banks. Many improved varieties, some thornless, are in cultivation. Wood reddish color.

P. Alleghaniénsis.

3. **Prùnus Alleghaniénsis**, Porter. (Alleghany Plum.) Leaves lanceolate to oblong-ovate, often long-acuminate, finely and sharply serrate, softly pubescent when young, smooth when old; fruit globose-ovoid, under ½ in., very dark purple, with a bloom; stone turgid, a shallow groove on one side and a broad, flat ridge on the other. A low, straggling bush, occasionally a tree, 3 to 15 ft. high. Mountains of Pennsylvania.

P. Chicàsa.

4. **Prùnus Chicàsa**, Michx. (Chicasaw Plum.) Leaves long, narrow, almost lanceolate, acute, finely serrate, thin. Flowers on short stalks. Fruit globular, ½ to 2/3 in. in diameter, thin-skinned, without bloom, yellowish-red, pleasant to taste. Stone globular, without sharp edges. A thorny shrub or small tree, 6 to 15 ft. high; wild in New Jersey, west and south, and often cultivated.

P. spinòsa.

5. **Prùnus spinòsa**, L. (Sloe. Blackthorn. Bullace Plum.) Leaves obovate-oblong to lance-oblong, sharply serrate, soon smooth; leaf-

stalk smooth; fruit small, globular, black, with a bloom; the stone rounded, acute at one edge; flesh greenish, astringent. A low tree with thorny branches; it is becoming naturalized along roadsides and waste places; from Europe. Var. *instititia* (Bullace Plum) is less thorny, and has the leafstalk and lower side of the leaves pubescent. [Pg 99]

P. doméstica.

6. **Prùnus doméstica**, L. (Common Garden Plum.) Leaves 1 to 3 in. long, oval or ovate-lanceolate, acute to obtuse. Flowers white, nearly solitary. Drupe globular, obovoid to ovoid, of many colors (black, white, etc.), covered with a rich glaucous bloom. A small tree, 10 to 20 ft. high, in cultivation everywhere for its fruit. Over a hundred varieties are named in the catalogues.

P. Pennsylvánica.

7. **Prùnus Pennsylvánica**, L. f. (Wild Red Cherry.) Leaves oblong-lanceolate, pointed, finely and sharply serrate, shining green, smooth on both sides. Flowers many in an umbel on long stems. Fruit round, light red, quite small, ¼ in. in diameter, sour. A small tree, 20 to 30 ft. high, in rocky woods; common north and extending southward along the Alleghanies to North Carolina.

P. àvium.

8. **Prùnus àvium**, L. (Bird-cherry or English Cherry.) Leaves oval-lanceolate, sharp-pointed, coarsely or doubly serrate. Flowers in sessile umbels, opening when the leaves appear. Fruit of various colors, somewhat heart-shaped. This is the Cherry tree, 30 to 50 ft. high, of which there are many named varieties usually cultivated for the fruit.

P. Cérasus.

9. **Prùnus Cérasus**, L. (Garden Red Cherry. Morello Cherry.) Leaves obovate and lance-ovate, serrate, on slender spreading branches. Flowers rather large. Fruit globular, bright red to dark purple, very sour; in sessile umbels. A small, round-headed tree, 10 to 30 ft. high, often cultivated. The preceding species and this one are the parents of most of the Cherry trees in cultivation.

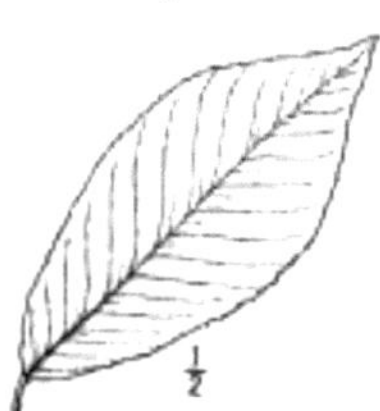

P. serótina.

10. **Prùnus serótina**, Ehrh. (Wild Black Cherry.) Leaves oblong or lance-oblong, thickish, smooth, usually taper-pointed, serrate, with incurved, short, thick teeth. Flowers in long racemes. June. [Pg 100] Fruit as large as peas, purple-black, bitter; ripe in autumn. A fine tree, 15 to 60 ft. high, with reddish-brown branches. Wood reddish and valuable for cabinet-work. Common in woodlands and along fences.

P. Virginiàna.

11. **Prùnus Virginiàna**, L. (Choke-cherry.) Leaves thin, oval-oblong or obovate, abruptly pointed, very sharply, often doubly serrate, with slender teeth. Racemes of flowers and fruit short and close. Fruit dark crimson, stone smooth. Flowers in May; fruit ripe in August; not edible till fully ripe. A tall shrub, sometimes a tree, with grayish bark. River-banks, common especially northward.

P. Pàdus.

12. **Prùnus Pàdus**, L. (Small Bird-cherry.) Like Prunus Virginiana, excepting that the racemes are longer and drooping, and the stone is roughened. Occasionally planted for ornament.

Genus **37. PỲRUS.**

Trees and shrubs, with alternate, stipulate, simple, or pinnately compound leaves. Flowers conspicuous, white to pink, apple-blossom-shaped (5 petals); in spring. Fruit a fleshy pome, with the cells formed by papery or cartilaginous membranes within juicy flesh.

* Leaves deeply pinnatifid or fully pinnate (Mountain Ashes) (**A.**)

    A. Leaf deeply pinnatifid, sometimes fully divided at the base.   6.

    A. Leaf once-pinnate throughout. (**B.**)

      B. Leaf-buds pointed, smooth and somewhat glutinous   7.

      B. Leaf-buds more or less hairy   8, 9.

* Leaves simple and not pinnatifid. (**C.**)

    C. Leaves entire; fruit solitary (Quinces)   5.

    C. Leaves serrate; fruit clustered. (**D.**)

      D. Fruit large, sunken at both ends (Apples)   1. [Pg 101]

      D. Fruit small (½-1 in.), sour, much sunken at the stem end and but little at the other (Crab-apples). (**E.**)

        E. Leaves very narrow; fruit ½ in.   2.

        E. Leaves broad; fruit 1 in.   3.

      D. Fruit usually obovate, not sunken at the stem end (Pears).   4.

P. Màlus.

1. **Pỳrus Màlus**, L. (Common Apple-tree.) Leaves simple, ovate, evenly crenate or serrate, smooth on the upper surface and woolly on the lower. Flowers large (1 in.), white, tinged with pink, in small corymbs. May. Fruit large, sunken at both ends, especially at base;

ripe from August to October, according to variety. A flat-topped tree, 20 to 40 ft. high, cultivated in hundreds of named varieties; from Europe.

P. angustifòlia.

2. **Pỳrus angustifòlia**, Ait. (Narrow-leaved Crab-apple.) Leaves simple, lanceolate or oblong, often acute at base, mostly serrate, smooth. Flowers large (2/3 in.), rose-colored, fragrant, in small, simple, umbel-like clusters. Fruit very sour, small (½ in.). Twigs lead-colored and speckled. A small tree, 12 to 20 ft. high. Pennsylvania and southward.

P. coronària.

3. **Pỳrus coronària**, L. (American or Garland Crab-apple.) Leaves simple, ovate, often rather heart-shaped, cut-serrate, often 3-lobed, soon smooth. Flowers large (¾ in.), few, in a cluster, rose-colored, very fragrant. Fruit very sour and astringent, flattened, broad, 1 in. or more in diameter, yellowish green. Small tree, 10 to 25 ft. high; New York, west and south, also frequently cultivated.

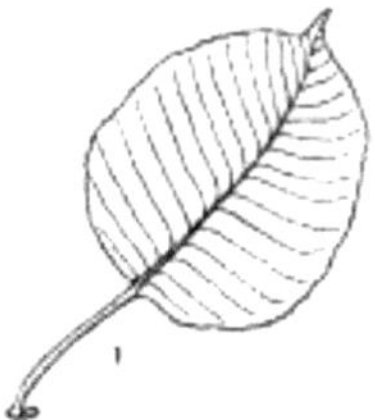

P. commùnis.

4. **Pỳrus commùnis**, L. (Common Pear-tree.) Leaves simple, ovate, serrate, smooth on both sides, at least when mature. Flowers large (over 1 in.), white, with purple anthers. April and May. [Pg 102] Fruit large, usually obovate and mainly sunken at the large end; ripe July to October, according to the variety. A pyramidal-shaped tree, 30 to 70 ft. high, with smooth bark and often somewhat thorny branches. Of several hundred named varieties, native to Europe. Cultivated for its fruit. Wood slightly tinged with red; strong, and of fine grain.

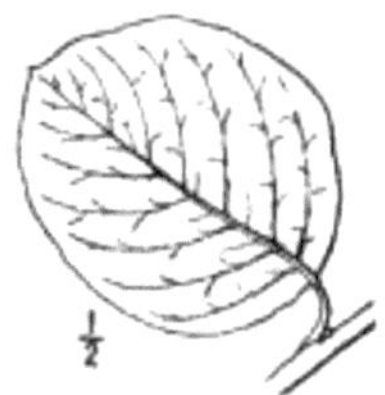

P. vulgàris.

5. **Pỳrus vulgàris.** (Quince. Common Quince-tree.) Leaves ovate, obtuse at base, entire, hairy beneath. Flowers solitary, large, 1 in., white or pale rose-color. Fruit large, hard, orange-yellow, of peculiar sour flavor; seeds mucilaginous; ripens in October. A low tree, 10 to 20 ft. high, with a crooked stem and rambling branches; from Europe. Several varieties in cultivation.

P. pinnatífida.

6. **Pỳrus pinnatífida**, Ehrh. (Oak-leaved Mountain-ash.) Leaves pinnately cleft and often fully pinnate at base, hairy beneath. Pome globose, ¼ in., scarlet, ripe in autumn. A cultivated tree, 20 to 30 ft. high; from Europe.

P. Americàna.

7. **Pỳrus Americàna**, DC. (American Mountain-ash.) Leaflets 13 to 15, lanceolate, bright green, nearly smooth, taper-pointed, sharply serrate with pointed teeth. Leaf-buds pointed, glabrous and somewhat glutinous. Flowers white, 1/3 in., in large, flat, compound cymes. In June. Fruit berry-like pomes, the size of small peas, bright scarlet when ripe in September, and hanging on the tree till winter. A tall shrub or tree, 15 to 30 ft. high, in swamps and mountain woods; more abundant northward. Often cultivated for the showy clusters of berries in autumn.

P. sambucifòlia.

8. **Pỳrus sambucifòlia**, Cham. & Schlecht. (Elder-leaved Mountain-ash.) Leaflets oblong, oval or lance-ovate, obtuse [Pg 103] (sometimes abruptly sharp-pointed), usually doubly serrate with rather spreading teeth, generally pale beneath. Leaf-buds somewhat hairy. Flowers and berries larger, but in smaller clusters, than the preceding species. The berries globose when ripe, 1/3 in. broad, bright red. This species, much like Pyrus Americana, is found wild in northern New England and westward.

P. aucupària.

9. **Pỳrus aucupària**, Gaertn. (European Mountain-ash, or Rowan-tree.) Much like Pyrus Americana, but the leaflets are paler and more obtuse, with their lower surface downy. Leaf-buds blunter and densely covered with hairs. Flowers larger, ½ in. or more in diameter. Fruit also much larger, sometimes nearly ½ in. in diameter. Beautiful tree, 20 to 30 ft. high, often cultivated.

Genus **38. CRATÆGUS.**

Thorny shrubs or small trees with simple, alternate, serrate, doubly serrate or lobed leaves. Flowers cherry-like blossoms, usually white in color and growing in corymbs, generally on the ends of side shoots; in spring. Fruit a berry or drupe with 1 to 5 bony stones, tipped with the 5 persistent calyx-teeth; ripe in autumn.

* Calyx, stipules, bracts, etc., often glandular. (**A.**)

  A. Flowers and fruit often over 6 in a cluster. (**B.**)

    B. Leaves usually abrupt at base   1.

    B. Leaves usually attenuate at base   2.

  A. Flowers and fruit few, 1 to 6 in a cluster   10.

* Calyx, etc., without glands (No. 4 has glandular teeth to the calyx); flowers many in a cluster. (**C.**)

C. Leaves more or less tapering at base. (**D.**)

  D. Leaves generally lobed; cultivated, rarely escaped   3.

  D. Leaves rarely lobed; native. (**E.**)

    E. Leaves small, shining, crenate at the end   5.

    E. Leaves villous or pubescent, at least when young   9.

    E. Leaves smooth or only downy at the axils, acutely serrate. South   7. [Pg 104]

C. Leaves usually abrupt at base, sometimes cordate. (**F.**)

  F. Leaves downy when young. (**G.**)

    G. Leaves usually lobed   4.

    G. Leaves rarely lobed; veins very prominent   8.

  F. Leaves quite smooth   6.

C. coccínea.

1. **Cratægus coccínea**, L. (Scarlet-fruited Thorn.) Leaves bright green, smooth, thin, roundish-ovate, sharply cut-toothed or lobed, on slender petioles. Branches reddish, villous-pubescent; spines stout, chestnut-brown. Flowers large, ½ to 2/3 in., many in a corymb, on glandular peduncles. May to June. Fruit scarlet, round or pear-shaped, ½ in.; ripe in September, with from 1 to 5 cells and seeds. Tall shrub or low tree, 10 to 25 ft. high, in hedges and woods; common from Canada to Florida.

Var. *mollis* has the shoots densely pubescent; leaves large, slender-petioled, cuneate, cordate or truncate at base, usually with acute narrow lobes, often rough above, and more or less densely pubescent beneath. Flowers large, 1 in.; fruit light scarlet with a light bloom, 1 in. broad.

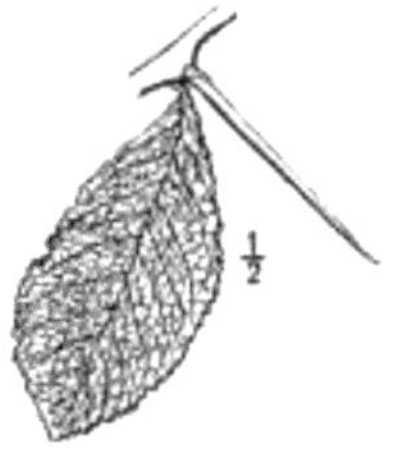

C. Crus-gálli.

2. **Cratǽgus Crus-gálli**, L. (Cockspur Thorn.) Leaves smooth, thick, shining above, wedge-obovate, finely serrate above the middle, with a short petiole. There are broad and narrow-leaved varieties. Flowers large and numerous, in lateral corymbs. May to June. Fruit globular, 1/3 in. broad, dull red; ripe in September and October. A small tree with a flat, bushy head, horizontal branches, and long, sharp thorns. Wild and common throughout, and often planted.

C. oxyacántha.

3. **Cratǽgus oxyacántha.** (English Hawthorn.) Leaves obovate, smooth, wedge-shaped at base, cut-lobed and [Pg 105] toothed above. No glands. Flowers medium-sized, ½ in., single or double, white, rose, or pink-red, numerous in corymbs. In spring. Fruit coral-red, 1/3 in.; ripe in autumn. A small tree or shrub, fine for lawn; from Europe; also escaped in some places.

C. apiifólia.

4. **Cratægus apiifólia**, Michx. (Parsley-leaved Thorn.) Leaves small, ovate, with a broad truncate or heart-shaped base, pinnatifid into 5 to 7 crowded, irregularly toothed lobes; white and soft-downy when young, smoothish when grown; petioles slender. Flowers medium-sized, ½ in., many in a corymb, white. May to June. Fruit small, 1/3 in., coral-red, ripe in autumn. A handsome, low (10 to 20 ft. high), spreading tree, with flexible branches and white-downy twigs. Virginia and south, in moist woods.

C. spathulàta.

5. **Cratægus spathulàta**, Michx. (Spatulate-leaved Thorn.) Leaves almost evergreen, thick, shining, spatulate, crenate toward the apex and nearly sessile, those on the young downy branches somewhat cut or lobed. Flowers small, ½ in., in large clusters. May. Fruit small, ¼ in., bright red; ripe in October. A small tree, 12 to 25 ft. high; Virginia and south.

C. cordàta.

6. **Cratægus cordàta**, Ait. (Washington Thorn.) Leaves broadly triangular-ovate, somewhat heart-shaped, thin, deep shining green, smooth, often 3- to 5-lobed and serrate, on slender petioles. Flowers small, 2/5 in., many in terminal corymbs, white. May, June. Fruit scarlet, about the size of peas; ripe in September. A compact, close-headed, small tree, 15 to 25 ft. high, with many slender thorns. Virginia, Kentucky, and southward. Sometimes planted in the North for hedges.

C. víridis.

7. **Cratægus víridis**, L. (Tall Hawthorn.) Leaves ovate to ovate-oblong, [Pg 106] or lanceolate, or oblong-obovate, mostly acute at both ends, on slender petioles; acutely serrate, often somewhat lobed and often downy in the axils. Flowers numerous, in large clusters. Fruit bright red, or orange, ovoid, small, ¼ in. broad. A small tree, 20 to 30 ft. high, with few large thorns or without thorns. Southern Illinois and Missouri, along the Mississippi and in the Southern States.

C. tomentòsa.

8. **Cratægus tomentòsa**, L. (Black or Pear Hawthorn.) Leaves downy-pubescent on the lower side (at least when young), thickish,

rather large, oval or ovate-oblong, sharply toothed and often cut-lobed below, abruptly narrowed into a margined petiole, the upper surface impressed along the main veins or ribs. Branches gray. Flowers ill-scented, many in a corymb. Fruit ½ in. long, obovate to globose, dull red. Shrub or tree, 10 to 30 ft. high, wild in western New York, west and south.

C. punctàta.

9. **Cratægus punctàta.** (Dotted-fruited Hawthorn.) Leaves rather small, mostly wedge-obovate, attenuate and entire below, unequally toothed above, rarely lobed, villous-pubescent, becoming smooth but dull, the veins prominent beneath and impressed above. Fruit globose, large, 1 in. broad, red to bright yellow; peduncles not glandular. Shrub to tree, 10 to 20 ft. high, with horizontal branches; Canada to Georgia.

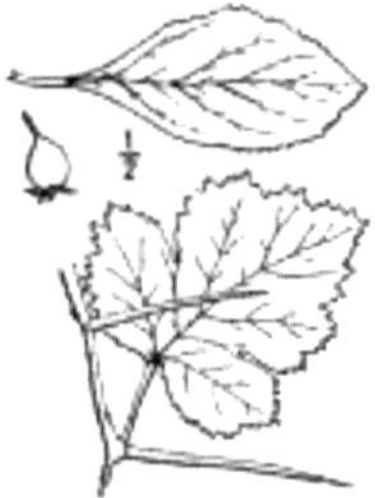

C. flàva.

10. **Cratægus flàva**, Ait. (Yellow or Summer Haw.) Leaves small, wedge-obovate, unequally toothed and cut above the middle; on short petioles; the teeth, stipules and petioles glandular. Flowers mostly solitary, white, large (¾ in). May. Fruit usually pear-shaped, quite large (¾ in. long), yellow or greenish-yellow, sometimes tinged or spotted with red, pleasant-flavored. Ripe in autumn. A low spreading tree, 15 to 20 ft. high. Virginia, south and west, in sandy soil.

Var. *pubescens* is downy-or villous-pubescent when young, and has thicker leaves and larger and redder fruit. [Pg 107]

Genus **39. AMELÁNCHIER.**

Small trees or shrubs with simple, deciduous, alternate, sharply serrate leaves; cherry-blossom-like, white flowers, in racemes at the end of the branches, before the leaves are fully expanded. Fruit a small apple-like pome; seeds 10 or less, in separate cartilaginous-coated cells.

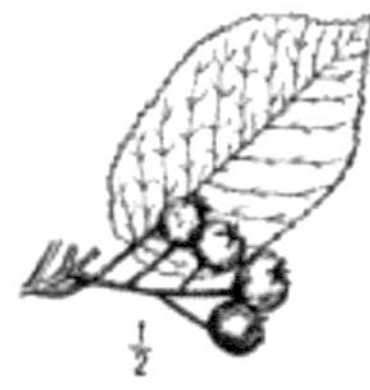

A. Canadénsis.

Amelánchier Canadénsis, Torr. & Gray. (Shad-bush. Service-berry.) A very variable species with many named varieties. The leaves, 1 to 3½ in. long, vary from narrow-oblong to roundish or cordate; bracts and stipules silky-ciliate. Flowers large, in drooping racemes, in early spring, with petals from 2 to 5 times as long as wide. Fruit globular, ½ in. broad, purplish, sweet, edible; ripe in June. It varies from a low shrub to a middle-sized tree, 5 to 30 ft. high.

Order **XIX. HAMAMELÍDEÆ.**

(Witch-Hazel Family.)

A small family of trees and shrubs represented in most countries.

Genus **40. HAMAMÈLIS.**

Tall shrubs, rarely tree-like, with alternate, straight-veined, 2-ranked, oval, wavy-margined leaves. Flowers conspicuous, yellow, 4-parted; blooming in the autumn while the leaves are dropping, and continuing in bloom through part of the winter. Fruit rounded capsules which do not ripen till the next summer.

H. Virginiána.

Hamamèlis Virginiána, L. (Witch-hazel.) The only species; 10 to 30 ft. high; rarely grows with a single trunk, but usually forms a slender, crooked-branched shrub. Flowers sessile, in small clusters of 3 to 4, in an involucre in the axils of the leaves. [Pg 108]

Genus **41. LIQUIDÁMBAR.**

Trees with alternate, simple, palmately cleft leaves. Flowers inconspicuous; in spring. Fruit a large (1 in.), globular, long-stalked, dry, open, rough catkin, hanging on the tree through the winter.

L. Styracíflua.

Liquidámbar Styracíflua, L. (Sweet Gum. Bilsted.) Leaves rounded, deeply 5- to 7-cleft, star-shaped, dark green, smooth and shining, glandular-serrate. Twigs often covered with corky ridges. A large, beautiful tree, 30 to 70 ft. high, with deeply furrowed bark. Connecticut, west and south; abundant south of 40° N. Lat. Well worthy of more extensive cultivation than it has yet received.

Order **XX. LYTHRÀCEÆ.**

(Loosestrife Family.)

A small order of shrubs, herbs, or trees; mainly tropical.

Genus **42. PÙNICA.**

Leaves simple, usually opposite, deciduous; flowers scarlet, with 5 petals and numerous stamens; fruit a many-seeded berry.

P. granàtum.

Pùnica granàtum, L. (Pomegranate-tree.) Leaves opposite, lanceolate, smooth, entire; flowers large, both calyx and corolla scarlet and very ornamental; the fruit as large as an orange, fine-flavored. A tree-shaped plant, growing to the height of 20 ft. in the Southern States. If given some protection, it can be grown as far north as Washington. It has been cultivated from the earliest times, and is probably a native of western Asia. [Pg 109]

Genus **43. LAGERSTRŒMIA.**

Flowers with 6 long-clawed petals inserted on the broadly spreading calyx; fruit 3- to 6-celled pods with many winged seeds.

L. Índica.

Lagerstrœmia Índica, L. (Crape-myrtle.) Leaves roundish-ovate, thick, smooth, short-petiolate; branches winged; flowers in terminal clusters with large, delicately crisped, long-stemmed petals of pink, purple, and other colors. A beautiful small tree, or usually a shrub, from India; often cultivated in the North in conservatories; hardy as far north as Washington.

Order **XXI. ARALIÀCEÆ.** (Ginseng Family.)

A small order of herbs, shrubs, and trees, here represented by the following genus:

Genus **44. ARÀLIA.**

Herbs, shrubs, or trees, with pinnately or palmately compound leaves; here including Acanthopanax with palmately cleft leaves. Flowers whitish or greenish, in umbels, often forming large panicles. Fruit small, berry-like, several-celled, several-seeded.

* Leaves 2 to 3 times odd-pinnate (Aralia proper) 1, 2.

* Leaves simple, palmately cleft (Acanthopanax) 3.

A. spinòsa.

1. **Aràlia spinòsa**, L. (Angelica-tree. Hercules'-Club.) Leaves large, crowded at the summit of the stem, [Pg 110] twice or sometimes thrice odd-pinnate, usually prickly, with sessile, ovate, acuminate, deeply serrate leaflets, glaucous beneath. Large panicles of small whitish flowers in umbels, with involucres of few leaves. Berry small, ¼ in., 5-ribbed, crowned with the remains of the calyx. A tree-like plant, 8 to 12 ft. high, or in the Gulf States 30 ft. high, with the stem covered with numerous prickles. Usually dies to the ground after flowering. Wild in damp woods, Pennsylvania and south, and cultivated in the North.

144

A. Chinénsis.

2. **Aràlia Chinénsis.** Leaves more or less fully twice-pinnate; leaflets ovate-oblong, oblique at base, acuminate, sharply serrate, hairy. Flowers and fruit in large, branching, hairy panicles; thorns few, straight. A small tree, 10 to 15 ft. high; occasionally cultivated; from China.

A. Maximowíczii.

3. **Aràlia (Acanthópanax) Maximowíczii.** Leaves long-petioled, simple, thick, palmately cleft, with 7 serrate lobes; old leaves smooth, the young with woolly bases. Panicles of flowers and fruit terminal; the berries striated. Tree-trunk usually quite prickly. This species is said to grow 50 ft. high in Japan. It has been recently introduced, and proves perfectly hardy in Massachusetts.

Order **XXII. CORNÀCEÆ.** (Dogwood Family.)

A small order of shrubs and trees (rarely herbs) of temperate regions.

Genus **45. CÓRNUS.**

Small trees or shrubs (one species an herb) with simple, entire, curved-veined, and (except in one species) op [Pg 111] posite leaves. The curved parallel ribs of the leaves in all the species are quite peculiar and readily recognized. Flowers small, of 4 petals, in some species rendered very conspicuous by large bracts. Fruit small, usually bright-colored drupes in clusters; ripe from August to October. There are but 3 species that grow at all tree-like.

* Leaves opposite. (**A.**)

    A. Fruit in close head-like clusters, red when ripe   1.

    A. Fruit in open clusters. (**B.**)

        B. Branches bright red; fruit white   2.

        B. Branches brownish; fruit bright red   3.

* Leaves alternate; fruit blue   4.

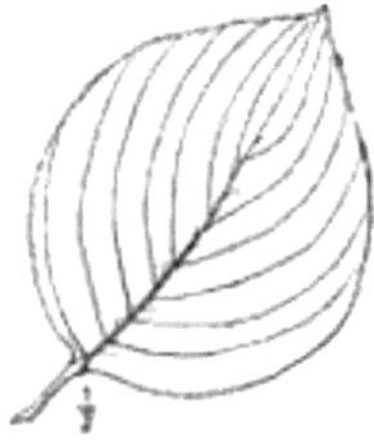

C. flórida.

1. **Córnus flórida**, L. (Flowering Dogwood.) Leaves ovate, pointed, acutish at base. Flowers in a head surrounded by 4 white bracts, making the whole cluster look like a single large flower 3 in. broad. Abundant in May and June. Fruit a small, bright red drupe with a single 2-seeded nut. Ripe in August. A large shrub or low tree 15 to 40 ft. high, with broad, roundish head. Common on high ground throughout, and one of the finest small trees in cultivation. A variety with the bracts quite red is also cultivated.

146

C. álba.

2. **Córnus álba**, L. (Siberian Red-stemmed Cornel.) Leaves broadly ovate, acute, densely pubescent beneath; drupes white; branches recurved, bright red, rendering the plant a conspicuous object in the winter. A shrub rather than a tree, cultivated from Siberia; hardy throughout.

C. máscula.

3. **Córnus máscula**, Dur. (Cornelian Cherry.) Leaves opposite, oval-acuminate, rather pubescent on both surfaces. Flowers small, yellow, in umbels from a 4-leaved involucre, blooming before the leaves are out in spring. Fruit oval, ½ in. long, cornelian-colored, ripe in autumn, rather sweet, used in confectionery. A large shrub or low tree, 8 to 15 [Pg 112] ft. high, with hard, tough, flexible wood, sometimes cultivated for its early flowers and late, beautiful fruit.

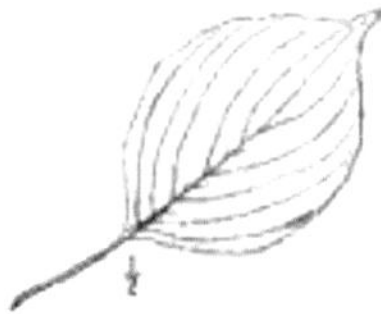

C. alternifòlia.

4. **Córnus alternifòlia**, L. f. (Alternate-leaved Cornel.) Leaves alternate, clustered at the ends of the branches, ovate or oval-

acuminate, tapering at base, whitish with minute pubescence beneath. Cymes of flowers and fruit broad and open. Fruit deep blue on reddish stalks. Shrub, though occasionally tree-like, 8 to 25 ft. high; on hillsides throughout; rarely cultivated.

Genus **46. NÝSSA.**

Trees with deciduous, alternate, exstipulate, usually entire leaves, mostly acute at both ends. Flowers somewhat diœcious, i.e. staminate and pistillate flowers on separate trees. The staminate flowers are quite conspicuous because so densely clustered. April and May. Fruit on but a portion of the trees, consisting of one or two small (¼ to ½ in.), drupes in the axils of the leaves. Stone roughened with grooves. Ripe in autumn.

* Fruit usually clustered	1, 2.
* Fruit solitary	3.

N. sylvática.

1. **Nýssa sylvática**, Marsh. (Pepperidge. Black or Sour Gum.) Leaves oval to obovate, pointed, entire (sometimes angulate-toothed beyond the middle), rather thick, shining above when old, 2 to 5 in. long. The leaves are crowded near the ends of the branches and flattened so as to appear 2-ranked, like the Beech; turning bright crimson in the autumn. Fruit ovoid, bluish-black, about ½ in. long, sour. Medium-sized [Pg 113] tree with mainly an excurrent trunk and horizontal branches. Wood firm, close-grained and hard to split. Rich soil, latitude of Albany and southward. Difficult to transplant, so it is rarely cultivated.

2. **Nýssa biflòra**, Walt. (Sour Gum.) Leaves 1 to 3 in. long, smaller than in N. sylvatica; fertile flowers and fruit 1 to 3, in the axils; stone decidedly flattened and more strongly furrowed. New Jersey to Tennessee and southward. Too nearly like the last to need a draw-

ing. All the species of Nyssa may have the margin of the leaves somewhat angulated, as shown in the next.

N. uniflòra.

3. **Nýssa uniflòra**, Wang. (Large Tupelo.) Leaves much larger, 4 to 12 in. long, sometimes slightly cordate at base, entire or angularly toothed, downy beneath. Fruit solitary, oblong, blue, 1 in. or more in length. Wood soft, that of the roots light and spongy and used for corks. In water or wet swamps; Virginia, Kentucky, and southward.

Order **XXIII. CAPRIFOLIÀCEÆ.**

(Honeysuckle Family.)

Shrubs (rarely herb or tree-like plants) of temperate regions.

Genus **47. VIBÚRNUM.**

Shrubs or small trees with opposite, simple, petioled leaves. Flowers light-colored, small but in large, conspicuous, flat-topped clusters at the ends of the branches; blooming in early summer. Fruit small, 1-seeded drupes with flattened stones; ripe in autumn.

* Leaves distinctly palmately lobed   1.

* Leaves pinnately veined and not lobed. (**A.**)

    A. Coarsely dentated   2.

    A. Finely serrated. (**B.**)

        B. Leaves long-acuminated   3.

        B. Obtuse or slightly pointed   4.

[Pg 114]

V. Ópulus.

1. **Vibúrnum Ópulus**, L. (Cranberry-tree.) Leaves palmately veined and strongly 3-lobed, broadly wedge-shaped or truncate at base, the spreading lobes mostly toothed on the sides and entire in the notches; petiole with 2 glands at the apex. Fruit in peduncled clusters, light red and quite sour (whence the name "Cranberry-tree"). A nearly smooth, small tree or shrub, 4 to 12 ft. high; wild along streams, and cultivated under the name of Snowball-tree or Guelder Rose. In this variety the flowers have all become sterile and enlarged. **Vibúrnum acerifòlium** (Arrow-wood) has also lobed leaves, and is much more common. This species never forms a tree, and has dark-colored berries.

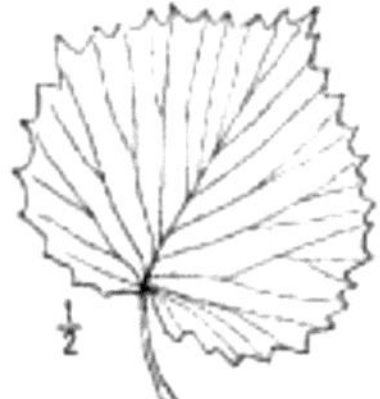

V. dentàtum.

2. **Vibúrnum dentàtum**, L. (Arrow-wood.) Leaves, pale green, broadly ovate, somewhat heart-shaped at base, coarsely and sharply dentated, strongly veined and often with hairy tufts in the axils; petioles rather long and slender. Fruit ¼ in. long, in peduncled clusters, blue or purple; a cross-section of the stone between kidney-and horseshoe-shaped. A shrub or small tree, 5 to 15 ft. high, with ash-colored bark; in wet places.

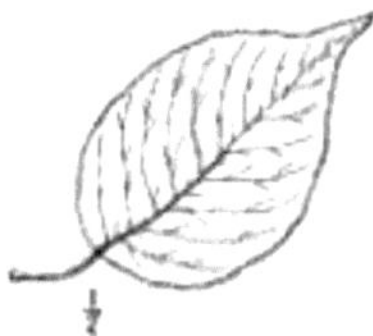

V. Lentàgo.

3. **Vibúrnum Lentàgo**, L. (Sweet Viburnum or Sheep-berry.) Leaves broad, ovate, long-pointed, 2 to 3 in. long, closely and sharply serrated; petioles long and with narrow, curled margins; entire plant smooth. Fruit in sessile clusters of 3 to 5 rays, oval, large, ½ in. long, blue-black, edible, sweet; ripe in autumn. A small tree, 10 to 30 ft. high; found wild throughout, in woods and along streams.

V. prunifòlium.

4. **Vibúrnum prunifòlium**, L. (Black Haw.) Leaves oval, obtuse or slightly pointed, 1 to 2 in. long, finely and sharply serrated. Blooming early, [Pg 115] May to June. Fruit oval, large (½ in. long), in sessile clusters of 3 to 5 rays, black or blue-black, sweet. A tall shrub or small tree, 6 to 12 ft. high; in dry soil or along streams; New York, south and west.

Genus **48. LONÍCERA.**

Leaves entire, opposite; corolla 5-lobed; berry several-seeded.

L. Tartárica.

Lonícera Tartárica. (Tartarian Honey-suckle.) Leaves deciduous, oval, heart-shaped; flowers in pairs, showy, pink to rose-red; in spring; berries formed of the two ovaries, bright red; ripe in summer. A shrub, often planted and occasionally trimmed to a tree-like form, and growing to the height of nearly 20 ft.

Order **XXIV. COMPÓSITÆ.**

This, the largest order of flowering plants, is made up almost exclusively of herbaceous plants, but contains one shrub or low tree which is hardy from Boston southward near the Atlantic coast.

Genus **49. BÁCCHARIS.**

Leaves simple, deciduous; heads of flowers small, many-flowered; receptacle naked; pappus of hairs.

B. halimifòlia.

Báccharis halimifòlia, L. (Groundsel-Tree.) Leaves obovate, wedge-shaped, crenately notched at end, light grayish in color, with whitish powder; branches angled; flowers white with a tint of purple, blooming in the autumn. A broad, loose-headed, light-colored bush rather than a tree, 8 to 15 ft. high; wild on sea-beaches, Massachusetts and south, and occasionally cultivated. The plant is diœcious; the fertile specimens are rendered quite conspicuous in autumn by their very long, white pappus. [Pg 116]

Order **XXV. ERICÀCEÆ.** (Heath Family.)

A large order, mainly of shrubs, though a few species are herbs, and fewer still are tall enough to be considered trees.

Genus **50. OXYDÉNDRUM.**

Trees with deciduous, alternate, oblong-lanceolate, pointed, serrate, sour-tasting leaves. Flowers small, in large panicles at the ends of the branches. In summer. Fruit small, dry capsules, with 5 cells and many seeds.

O. arbòreum.

Oxydéndrum arbòreum, DC. (Sorrel-tree. Sourwood.) Leaves in size and shape much like those of Peach trees. Flowers small, urn-shaped. Small-sized tree, 15 to 50 ft. high; wild in rich woods, Pennsylvania and southward, mainly in the mountains. Rare in cultivation, but very beautiful, especially in autumn, when its leaves are brilliantly colored, and the panicles of fruit still remain on the trees. It is perfectly hardy both at the Arnold Arboretum, Boston, and the Missouri Botanical Garden, St. Louis.

Genus **51. KÁLMIA.**

Evergreen shrubs with alternate, entire, thick, smooth leaves. Flowers large, beautiful, cup-shaped, in showy clusters. Fruit a small, 5-celled, many-seeded capsule.

K. latifòlia.

Kálmia latifòlia, L. (Mountain-laurel. Calico-bush.) The only species which grows at all tree-like has ovate-lanceolate or elliptical, smooth, petioled leaves, tapering at both ends and green on both sides. Flowers in terminal corymbs, clammy-pubescent, white to pink. June. Pod depressed, glandular. Shrub or small tree, 4 to 25 ft. high, with reddish twigs; wild in rocky hills and damp soils through out; occasionally planted. Wood very hard and close-grained. [Pg 117]

Genus **52. RHODODÉNDRON.**

Shrubs or low trees with usually alternate, entire leaves and showy flowers in umbel-like clusters from large, scaly-bracted, terminal buds. Fruit a dry 5-celled pod with many seeds.

R. máximum.

Rhododéndron máximum, L. (Great Laurel.) Leaves thick, 4 to 10 in. long, elliptical-oblong or lance-oblong, acute, narrowed toward the base, very smooth, with somewhat revolute margins. Flowers large (1 in.), with an irregular bell-shaped corolla and sticky stems, in large clusters, white or slightly pinkish with yellowish dots. July. Evergreen shrub or tree, 6 to 20 ft. high, throughout the region, especially in damp swamps in the Alleghany Mountains; occasionally cultivated.

Genus 53. CLÈTHRA.

Shrubs or trees with alternate, simple, deciduous, exstipulate, serrate leaves. Flowers (July and August) conspicuous, white, in elongated terminal racemes which are covered with a whitish powder. Fruit 3-celled pods with many seeds, covered by the calyx.

* Leaves thin, large, 3 to 7 in. long, pale beneath   1.

* Leaves thickish, smaller, green both sides   2.

C. acuminàta.

1. **Clèthra acuminàta**, Michx. (Acuminate-leaved Clethra. Sweet Pepper-bush.) Leaves 3 to 7 in. long, oval to oblong, pointed, thin, abruptly acute at base, finely serrate, on slender petioles, smooth above and glaucous below. Racemes drooping, of sweet-scented flowers, with the bracts longer than the flowers. Filaments and pod hairy. A small tree or shrub, 10 to 20 ft. high, in the Alleghanies, Virginia, and south. Not often in cultivation, but well worthy of it. [Pg 118]

C. alnifòlia.

2. **Clèthra alnifòlia**, L. (Common Sweet Pepper-bush.) Leaves wedge-obovate, sharply serrate near the apex, entire near the base, straight-veined, smooth, green on both sides. Racemes erect, often compound, with bracts shorter than the flowers and with smooth filaments. This is a shrub rather than a tree; abundant in wet places east of the Alleghanies. Occasionally cultivated for its sweet-scented flowers.

Order **XXVI. SAPOTÀCEÆ.**

(Sapodilla Family.)

A small order, mainly of tropical plants, here including one genus found only in the southern part of our range.

Genus **54. BUMÈLIA.**

Leaves simple, alternate, entire, sub-evergreen, exstipulate; branches often spiny. Flowers small, whitish, usually crowded in fascicles. Fruit a black cherry-like drupe with a 2- to 3-celled nut. Shrubs and trees of the Southern States. Two species (although hardly trees) are found far enough north to be included in this work.

* Leaves rusty-woolly beneath   1.

* Leaves smooth or slightly silky beneath 2.

B. lanuginòsa.

1. **Bumèlia lanuginòsa**, Pers. (Woolly-leaved Buckthorn.) Leaves oblong-obovate, obtuse, entire, smooth above and rusty-woolly beneath, but not silky; spiny, with downy branchlets. Clusters 6- to 12-flowered, pubescent; flowers greenish-yellow. Fruit globular and quite large (½ in.), black, edible. A small tree, 10 to 40 ft. high, of the woods of southern Illinois and southward. With slight protection it can be cultivated in Massachusetts. [Pg 119]

B. lycioìdes.

2. **Bumèlia lycioìdes**, Pers. (Southern Buckthorn.) Leaves 2 to 4 in. long, oval-lanceolate, usually bluntish with a tapering base and entire margin, deciduous, a little silky beneath when young. Clusters densely many-flowered (20 to 30); flowers small (1/6 in.), smooth, greenish-white. May, June. A spiny shrub or tree, 10 to 25 ft. high, in moist ground, Virginia, west and south. About as hardy as the preceding species.

Order **XXVII. EBENÀCEÆ.** (Ebony Family.)

A small order of mostly tropical trees and shrubs.

Genus **55. DIOSPÝROS.**

Trees or shrubs with alternate, simple, entire, feather-veined leaves. Flowers small, inconspicuous, mostly diœcious. Fruit a globose berry with the 5-lobed thick calyx at the base, and with 8 to 12, occasionally 1 to 5, rather large seeds; ripe after frost.

D. Virginiàna.

D. Lòtus and D. Kàki.

Diospýros Virginiàna, L. (Common Persimmon.) Leaves 4 to 6 in. long, ovate-oblong, acuminate, rather thick, smooth, dark, shining above, a little pale beneath. Bark dark-colored and deeply furrowed in a netted manner with rather small meshes. Flowers yellowish, rather small, somewhat diœcious; the staminate ones urn-shaped with mouth nearly closed; the pistillate ones more open. June. Fruit large, 1 in.; very astringent when young, yellow and pleasant-tasting after frost. A handsome, ornamental tree, 20 to 60 ft. high, with very hard, dark-colored wood and bright foliage. Southern New England to Illinois [Pg 120] and south; also cultivated. **Di-**

ospỳros Lòtus (Date-plum), with leaves very dark green above, much paler and downy beneath, and fruit much smaller (2/3 in.), and **Diospỳros Kàki** (Japan Persimmon), with large, leathery, shining leaves and very large fruit (2 in.), are successfully cultivated from Washington, D. C., southward. The under leaf represents D. Lotus, the upper one a small specimen of D. Kaki.

Order **XXVIII. STYRACÀCEÆ.**

(Storax Family.)

A small order of shrubs and trees, mostly of warm countries.

Genus **56. STỲRAX.**

Shrubs or small trees with commonly deciduous leaves, and axillary, or racemed, white, showy flowers on drooping stems. Pubescence scurfy or stellate; fruit a globular dry drupe, its base covered with the persistent calyx, forming a 1- to 3-seeded nut.

S. Americàna.

1. **Stỳrax Americàna**, Lam. (American Storax.) Shrub or small tree (4 to 10 ft.), with oblong, alternate leaves acute at both ends, 1 to 3 inches long, smooth or very nearly so; fruit ½ in. long, in racemes of 3-4. Wild along streams, Virginia and south; occasionally cultivated, and probably hardy throughout.

S. Japónica.

2. **Stỳrax Japónica**, Sieb. (Japan Storax.) Leaves alternate, membranaceous, ovate to ovate-lanceolate, serrate or crenate, ½ to 3 in. long, smooth or with short [Pg 121] stellate hairs; flowers and fruit in long racemes. A beautiful low tree, 6 to 12 ft. high; from Japan. Hardy as far north as Philadelphia, but needing a little protection in Massachusetts and Missouri.

Genus **57. PTEROSTỲRAX.**

Similar to Styrax, but with the fruit in panicles, 5-winged, conical, and crowned with the persistent base of the style.

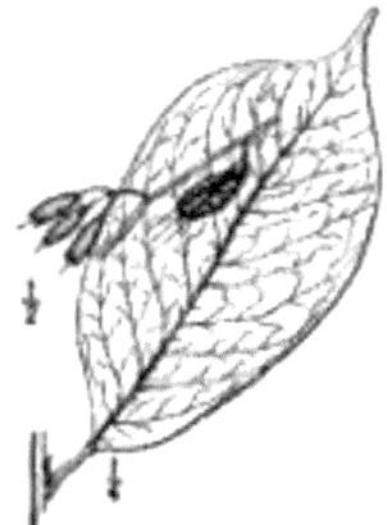

P. corymbòsum.

Pterostỳrax corymbòsum, Sieb. Leaves deciduous, 2 to 5 in. long, feather-veined, petioled, ovate, rarely cordate at base, sharply serrate, with stellate hairs. Shrub or small tree, 10 to 12 ft. high, cultivated from Japan; with ashy-gray bark, and white flowers turning yellowish or purplish with age; blooming in May, fruit ripe in August. Not perfectly hardy in Massachusetts.

Genus **58. HALÈSIA.**

Small trees or shrubs with alternate, simple, deciduous, serrate leaves. Flowers large, 1 in. long, conspicuous, white, hanging, bell-shaped, monopetalous, 4-lobed; blooming in spring. Fruit with a single, rough, elongated, bony nut surrounded by a 2- to 4-winged coat; ripe in autumn.

Wood light-colored, very hard and fine-grained.

H. díptera.

1. **Halèsia díptera, L.** (Two-winged Silverbell Tree.) Leaves large (4 to 5 in. long), ovate, acute, serrate, softly pubescent. Fruit with 2 conspicuous, broad wings, sometimes with 2 intermediate narrow ridges. A small tree or a large shrub, wild in the south, and cultivated as far north as New York City.

H. tetráptera.

2. **Halèsia tetráptera, L.** (Four-winged Silverbell Tree.) Leaves smaller (2 to 4 in.), oblong-ovate, finely serrate. Fruit smaller, with [Pg 122] 4 nearly equal wings. A small, beautiful tree, 10 to 30 ft. high, more hardy than Halesia diptera, and therefore cultivated occasionally throughout. Wild in Virginia and south.

Genus 59. SÝMPLOCOS.

Shrubs or small trees, with leaves furnishing a yellow dye.

S. tinctòria.

Sýmplocos tinctòria, L'Her. (Horse-sugar. Sweetleaf.) Leaves simple, alternate, thick, 3 to 5 in. long, elongate-oblong, acuminate, nearly entire, almost persistent, pale beneath, with minute pubescence, sweet-tasting. Flowers 6 to 14, in close-bracted, axillary clusters, 5-parted, sweet-scented, yellow; in early spring. Fruit a dry drupe, ovoid, ½ in. long. A shrub or small tree, 10 to 20 ft. high. Delaware and south.

Order **XXIX. OLEÀCEÆ.** (Olive Family.)

An order of trees and shrubs, mainly of temperate regions.

Genus **60. FRÁXINUS.**

Trees with petioled, opposite, odd-pinnate leaves (one cultivated variety has simple leaves). Flowers often inconspicuous, in large panicles before the leaves in spring. Fruit single-winged at one end (samara or key-fruit), in large clusters; ripe in autumn. Some trees, owing to the flowers being staminate, produce no fruit. Wood light-colored, tough, very distinctly marked by the annual layers. The leaves appear late in the spring, and fall early in the autumn.

* Flowers with white corolla; a cultivated small tree   8.

* Flowers with no corolla. (**A.**)

  A. Leaves pinnate; leaflets petiolate; calyx small, persistent on the fruit. (**B.**)

    B. Fruit broad-winged, ¾ in. wide. South   5.

    B. Wings much narrower. (**C.**) [Pg 123]

      C. Branchlets round and pubescent   2.

      C. Branchlets round and smooth. (**D.**)

        D. Leaflets nearly entire   1.

        D. Leaflets serrate near tip, entire below   3.

      C. Branchlets, on vigorous growths, square   4.

  A. Leaves pinnate; leaflets sessile; no calyx. (**E.**)

    E. Native; wing of fruit rounded at tip   6.

    E. Cultivated from Europe; wing notched at tip   7.

  A. Leaves simple; variety under   7.

F. Americàna.

1. **Fráxinus Americàna**, L. (White Ash.) Leaflets 7 to 9 (usually 7), stalked, ovate or lance-oblong, pointed, shining above, pale and either smooth or pubescent beneath, somewhat toothed or entire. Flowers almost always diœcious (May), thus the fruit is found on but a portion of the trees. The fruit (August to September) terete and marginless below, abruptly dilated into the wing, which is 2 to 3 times as long as the terete portion; entire fruit about 1½ in. long. A common large forest-tree, 60 to 80 ft. high, with gray, furrowed bark, smooth, grayish-green branchlets, and rusty-colored buds. Extensively cultivated.

F. pubéscens.

2. **Fráxinus pubéscens**, Lam. (Red Ash.) Like the White Ash, but to be distinguished from it by the down on the young, green or olive-green twigs, and on the footstalks and lower surface of the leaves. Fruit acute, 2-edged at base, gradually dilated into the wings as in Fraxinus viridis. A smaller and more slender tree than the White Ash; growing in about the same localities, but rare west of the Alleghanies; heart-wood darker-colored.

F. víridis.

3. **Fráxinus víridis**, Michx. f. (Green Ash.) Smooth throughout; leaflets 5 to 9, bright green on both sides, ovate or oblong-lanceolate, often wedge-shaped at base and serrate above. Fruit acute and 2-edged or mar [Pg 124] gined at base and gradually spreading into an oblanceolate or linear-spatulate wing as in the Red Ash. Small to middle-sized trees (like the Red Ash), found throughout, but common westward.

F. quadrangulàta.

4. **Fráxinus quadrangulàta**, Michx. (Blue Ash.) Leaflets 7 to 9, short-stalked, oblong-ovate or lanceolate, pointed, sharply serrate, green on both sides. Fruit narrowly oblong, blunt, of the same width at both ends, or slightly narrowed at the base. A large tree, 60 to 80 ft. high, with smooth square twigs on the vigorous growths. Wisconsin to Ohio and Kentucky.

F. platycárpa.

5. **Fráxinus platycárpa**, Michx. (Water-ash.) Leaflets 5 to 7, 3 to 5 in. long, ovate or oblong, acute at both ends, short-stalked, slightly serrate. Branchlets terete, smooth to pubescent. Fruit broadly winged, ¾ in. wide, often 3-winged, tapering to the base. A medium-sized tree in deep river-swamps, Virginia and south.

F. sambucifòlia.

6. **Fráxinus sambucifòlia**, Lam. (Black Ash.) Leaflets 7 to 11, sessile, oblong-lanceolate, tapering to a point, serrate, obtuse or rounded at base, green and smooth on both sides; when young, with some rusty hairs along the midrib. Fruit without calyx at base and with wing all around the seed-bearing part, blunt at both ends. A slender tree, 40 to 70 ft. high, with dark-blue or black buds.

F. excélsior.

Var. monophýlla.

7. **Fràxinus excélsior**, L. (European Ash.) Leaflets 11 to 13 (in some culti [Pg 125] vated varieties reduced to 1 to 5), almost sessile, lanceolate-oblong, acuminate, serrate, wedge-shaped at base. Flowers naked, somewhat diœcious, and so the fruit does not form on all the trees. Keys linear-oblong, obtuse, obliquely notched at apex. This species in its very numerous varieties is common in cultivation. One of the most interesting is the Weeping Ash (var. *pendula*). The most remarkable is the one with simple, from pinnatifid to entire leaves (var. *monophylla*).

F. òrnus.

8. **Fráxinus òrnus.** (Flowering Ash.) Leaflets 7 to 9, lanceolate or elliptical, attenuated, serrated, entire at the stalked bases, villous or downy beneath. Flowers fringe-like, white, in large terminal drooping clusters, of 4 or 2 petals. May to June. Fruit small, lance-linear, obtuse, attenuate at each end. A small tree, 15 to 30 ft. high, planted in parks. Not hardy north of New York City without some protection.

Genus **61. OSMÁNTHUS.**

Shrub or small tree with opposite, thick, evergreen, nearly entire leaves. Flowers small, white, in panicles or corymbs in late spring. Fruit a spherical drupe, ½ in. long, with a 2-seeded stone; hanging on during the winter.

O. Americàna.

Osmánthus Americàna, L. (Devil-wood.) Leaves thick, evergreen, oblong-lanceolate, entire, acute, narrowed to a petiole, 4 to 5 in. [Pg 126] long. Flowers diœcious, very small. May. Fruit globular, about ½ in. in diameter, violet-purplish; ripe in autumn, and remaining on the tree through the winter. A small tree, 15 to 20 ft. high, from southern Virginia southward, in moist woods.

Genus **62. SYRÍNGA.**

Leaves simple, entire, opposite; flowers ornamental, in large, dense clusters. The Lilacs are all beautiful, but form mere shrubs, except the following:

S. Japónica.

Syrínga Japónica. (Japan Lilac. Giant Tree Lilac.) Leaves deciduous, opposite, oval to cordate, thick, dark green, glossy; flowers white, 4-parted, odorless, in very large, dense, erect, terminal clusters, blooming in summer; fruit dry 2-celled pods with 2 to 4 seeds. A magnificent small tree, 20 to 30 ft. high; from Japan; probably hardy throughout.

Genus **63. CHIONÁNTHUS.**

Low trees or shrubs with simple, deciduous, opposite, entire, thick, smooth, petioled leaves. Flowers 4-parted, with long, slender, delicate white lobes, drooping in clusters from the lower side of the branches and forming a fringe; in early summer. Fruit a purple drupe.

C. Virgínica.

Chionánthus Virgínica, L. (Fringe-tree). Leaves smooth, thickish, large (3 to 6 in. long), oval or obovate, entire. The leaves are occasionally somewhat alternate and thin; they resemble those of the Magnolia. Drupe ovoid, ¾ in. long, covered with a bloom. A beautiful small tree or shrub, 8 to 30 ft. high, wild along streams, southern Pennsylvania and southward, and generally cultivated north for its delicate fringe-like flowers. Hardy.

A variety (var. *angustifolia*) with long, narrow leaves is occasionally cultivated. [Pg 127]

Order **XXX. SCROPHULARIÀCEÆ.**

(Figwort Family.)

A large order of plants, almost entirely herbaceous; found in all climates; it includes one cultivated tree in this region.

Genus **64. PAULÒWNIA.**

Tree with opposite (sometimes in whorls of three), large, deciduous, palmately veined, heart-shaped leaves. Leaf-stem often hollow; minute cup-shaped glands, separated from one another, situated on many portions of the leaf, but quite abundant on the upper side at the branching of the veins. Flowers large, in immense panicles; in spring, before the leaves expand. Fruit a dry, ovate, pointed capsule, 1½ in. long, with innumerable flat-winged seeds; hanging on the tree throughout the winter.

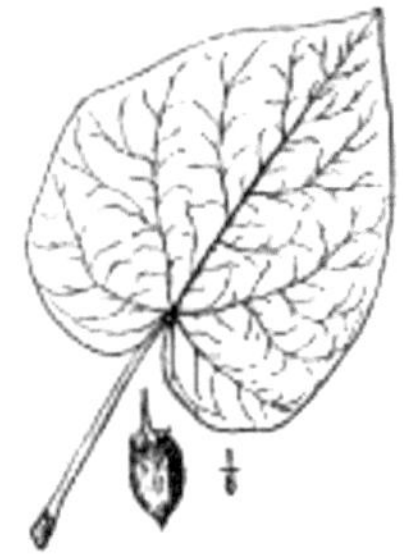

P. imperiàlis.

Paulòwnia imperiàlis, (Imperial Paulownia.) Leaves 7 to 14 in. long, sometimes somewhat lobed, usually very hairy beneath; 2 buds, almost hidden under the bark, above each other in the axil. Flowers purple, nearly 2 in. long, with a peculiar, thick, leather-like calyx. A broad flat-headed tree, of rapid growth when young. Cultivated; from Japan; and hardy throughout, but the flower-buds are winter-killed quite frequently north of New York City.

Order **XXXI. BIGNONIÀCEÆ.**

(Bignonia Family.)

An order of woody plants abundant in South America; here including one genus of trees: [Pg 128]

Genus **65. CATÁLPA.**

Trees or shrubs with large, simple, opposite (or whorled in threes), heart-shaped, pointed leaves. Flowers irregular, showy, in large panicles; blooming in June. Fruit long pods with many, winged seeds, hanging on till spring. Branches coarse and stiff. Wood light and close-grained.

* Flowers bright-spotted; wings of seeds narrowed   1.

* Flowers nearly pure white; wings of seeds broad   2.

C. bignonioìdes.

1. **Catálpa bignonioìdes**, Walt. (Indian Bean. Southern Catalpa.) The large heart-shaped leaf has connected scaly glands in the axils of the large veins on the lower side; usually entire though sometimes angulated, generally opposite though sometimes in whorls of threes, very downy beneath when young, 6 to 12 in. long. Flowers much spotted with yellow and purple, and with the lower lobe entire. Pod thin, 10 in. or more in length. A medium-sized, wide-spreading tree, 20 to 40 ft. high, of rapid growth, with soft, light wood and thin bark; wild in the Southern States, and extensively cultivated as far north as Albany.

C. speciósa.

2. **Catálpa speciósa**, Warder. (Indian Bean. Western Catalpa.) Leaves large (5 to 12 in. long), heart-shaped, long-pointed. Flowers 2 in. long, nearly white, faintly spotted, the lower lobes somewhat notched. Pod thick. A large, tall tree, 40 to 60 ft. high, with thick bark; wild in low, rich woodlands, southern Indiana, south and west. [Pg 129]

C. Kæmpferi.

Catálpa Kæmpferi and **Catálpa Búngei** are dwarf forms from Japan, the latter growing to the height of from 4 to 8 ft., and the former rarely reaching the height of 18 ft. The leaf of C. Kæmpferi is figured. It is more apt to have its margin angulated, though all the species occasionally have angulated leaves.

Order **XXXII. VERBENACEÆ.**

Herbs, shrubs, rarely small trees, with opposite leaves, irregular flowers and dry 2- to 4-celled fruits.

Genus **66. CLERODÉNDRON.**

Shrubby trees or climbing shrubs with opposite or whorled, usually entire leaves; flowers with an almost regular, 5-parted corolla surrounded by a bell-shaped calyx; fruit drupe-like, with 4 seeds.

C. trichótomum.

Clerodéndron trichótomum, Thunb. (Fate-tree.) Leaves opposite, long-petioled, cordate, thin, entire, glandular-dotted above, very veiny; lower leaves largest and three-lobed, the upper ovate, long-pointed, all 3-ribbed. Flowers in large, terminal clusters; fruit with juicy pulp covering the 4 seeds. A small tree from Japan; hardy at Washington and south. The figure represents one of the upper leaves.

Genus 67. VÍTEX.

Shrubs or low trees with opposite, usually palmate leaves, panicled clusters of flowers and drupe-like fruit. [Pg 130]

V. Agnus-cástus.

Vítex Agnus-cástus, L. (Chaste-tree.) Leaves long-petioled, palmate, with 5 to 7 lanceolate, acute, nearly entire leaflets, whitened beneath; with an aromatic though unpleasant odor. Branches obtusely 4-sided, hairy; flowers pale lilac, in interrupted panicles, agreeably sweet-scented in late summer. Shrub or small tree, 5 to 10 ft. high, cultivated from southern Europe; hardy at Washington and south. If cultivated further north, it needs protection, at least when young.

Order XXXIII. LAURÀCEÆ. (Laurel Family.)

An order of aromatic trees and shrubs, chiefly tropical.

Genus 68. PÉRSEA.

Aromatic, evergreen trees with alternate, entire, feather-veined leaves. Flowers small, in small close panicles. Fruit small (½ in.) 1-seeded drupes.

P. Carolinénsis.

Pérsea Carolinénsis, Nees. (Red Bay.) Leaves 2 to 5 in. long, oblong, entire, covered with a fine down when young, soon smooth above. Flowers silky, in small rounded clusters on short stems. May. Fruit an ovate, pointed, 1-seeded, deep-blue drupe, ½ in. long, on a red stalk; ripe in autumn. Usually a small tree, 15 to 70 ft. high, wild in swamps, Delaware, Virginia, and south. Wood reddish, beautiful, hard, strong, durable.

Genus **69. SÁSSAFRAS.**

Aromatic trees or shrubs with alternate, simple, deciduous, often lobed leaves. Juice of bark and leaves [Pg 131] mucilaginous. Flowers yellowish-green, in clusters; blooming in early spring. Fruit a small bluish drupe on a thick reddish stem. Ripe in September. Twigs greenish-yellow.

S. officinàle.

Sássafras officinàle, Nees. (Sassafras.) Leaves very variable in form, ovate, entire, or some of them 2- to 3-lobed, soon smooth. Flowering as the leaves are putting forth. Tree 15 to 100 ft. high, common in rich woods. The aromatic fragrance is strongest in the bark of the roots. Wood reddish, rather hard and durable.

Genus **70. LÍNDERA.**

Shrubs with deciduous, alternate, aromatic leaves and small, yellow flowers in close clusters along the branches. Fruit a drupe on a not-thickened stalk.

L. Benzòin.

Líndera Benzòin, Blume. (Spice-bush. Benjamin-bush.) Leaves alternate, oblong-ovate, entire, pale beneath, very spicy in odor and taste; twigs green; leaf-buds scaly; drupes red, ripe in autumn. Flowers 4 to 5 together in sessile umbels; in early spring, before the leaves expand. Common in damp woods throughout.

Order **XXXIV. ELÆAGNÀCEÆ.**

(Oleaster Family.)

A small order of shrubs or small trees, with the leaves covered with silvery scurf.

Genus **71. ELÆÁGNUS**.

Leaves alternate, entire; flowers axillary, stemmed; fruit drupe-like with an 8-grooved stone. [Pg 132]

E. lóngipes.

Elæágnus lóngipes. (Silver-leaved Elæagnus.) Leaves almost evergreen, rather thick, ovate-oblong, rather blunt, entire, smooth and dark green above, but silvery below. Flowers inconspicuous. Fruit about ½ in. long, bright red, with silvery scales, very abundant and beautiful; ripe in July; juicy and edible, with a pungent flavor. Shrub from Japan; hardy throughout.

Genus **72. SHEPHÉRDIA.**

Small trees or shrubs with opposite, deciduous, entire, silvery-scaled leaves. Flowers very small, diœcious. Fruit small, berry-like, translucent, 1-seeded.

S. argéntea.

Shephérdia argéntea, Nutt. (Buffalo-berry. Rabbit-berry.) Leaves opposite, oblong-ovate, tapering at base, silvery on both sides, with small peltate scales. Branches often ending in sharp thorns. Fruit, scarlet berries the size of currants, forming continuous clusters on every branch and twig, but found only on the pistillate plants. They are juicy, somewhat sour, pleasant-tasting, and make excellent jelly; ripe in September. A small handsome tree, 5 to 20 ft. high, wild in the Rocky Mountains, and sometimes cultivated east. Its thorny-tipped branches make it a good hedge-plant. Hardy.

Order **XXXV. EUPHORBIÀCEÆ.**

(Spurge Family.)

A large order of mainly herbaceous and shrubby plants of warm countries, with usually milky juice.

Genus **73. BÚXUS.**

Shrubs or trees with opposite, evergreen, entire leaves and small flowers. The fruit 3-celled, 6-seeded pods. [Pg 133]

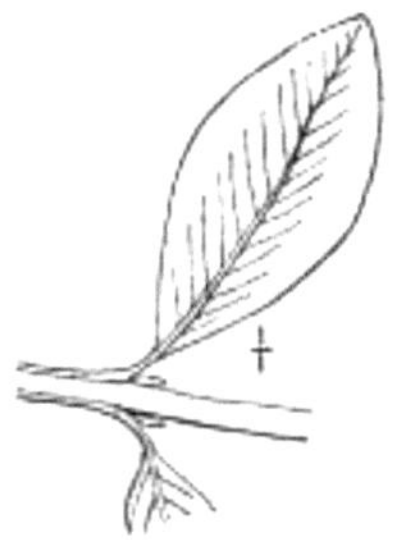

B. sempérvirens.

Búxus sempérvirens, L. (Boxwood.) Leaves ovate, smooth, dark green; leaf-stems hairy at edge. This plant is a native of Europe, and in its tree form furnishes the white wood used for wood-engraving.

Var. *subfruticosa* (dwarf boxwood) grows only a foot or two high, and is extensively used for edgings in gardens. The tree form is more rare in cultivation, and is of slow growth, but forms a round-topped tree.

Order **XXXVI. URTICÀCEÆ.** (Nettle Family.)

A large order of herbs, shrubs and trees, mainly tropical.

Genus **74. ÚLMUS.**

Tall umbrella-shaped trees with watery juice and alternate, 2-ranked, simple, deciduous, obliquely ovate to obliquely heart-shaped, strongly straight-veined, serrate leaves, harsh to the touch, often rough. Flowers insignificant, appearing before the leaves. Fruit a flattened, round-winged samara; ripe in the spring and dropping early from the trees. Bark rough with longitudinal ridges.

* Leaves very rough on the upper side. (**A.**)

A. Leaves 4 to 8 in. long; buds rusty-downy; inner bark very mucilaginous   1.

A. Leaves smaller; buds not downy; cultivated. (**B.**)

B. Wide-spreading tree; twigs drooping; fruit slightly notched   2.

B. Tree rather pyramidal; twigs not usually drooping; fruit deeply notched   3.

* Leaves not very rough on the upper side. (**C.**)

  C. Buds and branchlets pubescent; twigs often with corky ridges 4.

  C. Buds and branchlets free from hairs, or very nearly so. (**D.**)

    D. Twigs with corky wings 5.

    D. Twigs often with corky ridges; cultivated 2, 3.

    D. Branchlets never corky 6.

[Pg 134]

U. fúlva.

1. **Úlmus fúlva**, Michx. (Slippery or Red Elm.) Leaves large, 4 to 8 in., very rough above, ovate-oblong, taper-pointed, doubly serrate, soft-downy beneath; branchlets downy; inner bark very mucilaginous; leaves sweet-scented in drying; buds in spring soft and downy with rusty hairs. Fruit with a shallow notch in the wing not nearly reaching the rounded nut. A medium-sized tree, 45 to 60 ft. high, with tough and very durable reddish wood; wild in rich soils throughout.

U. montàna.

2. **Úlmus montàna**, Bauh. (Scotch or Witch Elm.) Leaves broad, obovate, abruptly pointed and doubly serrated. Fruit rounded, with a slightly notched wing, naked. Branches drooping at their extremity, their bark smooth and even. A medium-sized tree, 50 to 60 ft. high, with spreading or often drooping branches; extensively cultivated under a dozen different names, among the most peculiar being the White-margined (var. *alba marginata*), the Crisped-leaved (var. *crispa*), and the Weeping (var. *pendula*) Elms.

U. campéstris.

3. **Úlmus campéstris**, L. (English or Field Elm.) Leaves much smaller and of a darker color than the American Elm, obovate-oblong, abruptly sharp-pointed, doubly serrated, rough. Fruit smooth, with the wing deeply notched. A tall and beautiful cultivated tree, with the branches growing out from the trunk more abruptly than those of the American Elm, and thus forming a more pyramidal tree. A score of named varieties are in cultivation in this country, some with very corky bark, others with curled leaves, and still others with weeping branches.

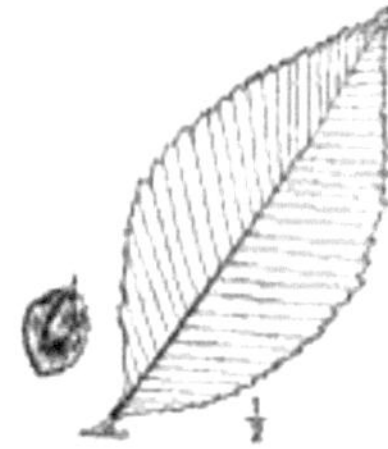

U. racemòsa.

4. **Úlmus racemòsa**, Thomas. (Cork or Rock Elm.) Leaves 2 to 4 in. long, obovate-oblong, abruptly pointed, often doubly ser [Pg 135] rated, with very straight veins; twigs and bud-scales downy-ciliate; branches often with corky ridges. Fruit large (½ in. or more long), with a deep notch; hairy. A large tree with fine-grained, heavy and

very tough wood. Southwest Vermont, west and south, southwest-ward to Missouri, on river-banks.

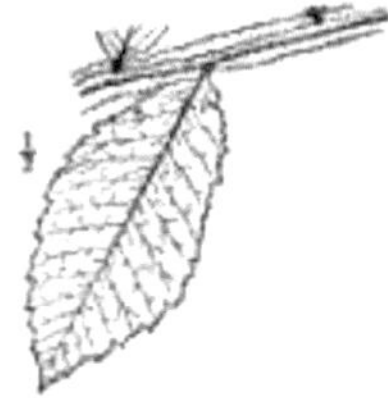

U. alàta.

5. **Úlmus alàta**, Michx. (Wahoo or Winged Elm.) Leaves small, 1 to 2 in. long, ovate-oblong or oblong-lanceolate, acute, thickish, downy beneath and nearly smooth above, sharply serrate. Bud-scales and branchlets nearly smooth. Notch in the wing of the fruit deep. A small tree, 30 to 40 ft. high, the branches having corky wings. Wild, Virginia, west and south; rarely cultivated.

U. Americàna.

6. **Úlmus Americàna**, L. (American or White Elm.) Leaves 2 to 4 in. long, obovate-oblong or oval, abruptly sharp-pointed, sharply and often doubly serrated, soft-pubescent beneath when young, soon quite smooth; buds and branchlets smooth. Fruit ½ in. long, its sharp points incurved and closing the deep notch; hairy only on the edges. A large ornamental tree, usually with spreading branches and drooping branchlets, forming a very wide-spreading top. Wild throughout in rich, moist soil; common in cultivation.

Genus **75. PLÁNERA.**

Trees or tall shrubs with alternate, simple, pointed, 2-ranked, feather-veined, toothed leaves. Flowers inconspicuous, with the leaves in spring. Fruit a small, nut- [Pg 136] like, scaly, globular drupe, ripe in autumn. Bark scaling off like that of the Sycamore.

P. aquática.

1. **Plánera aquática**, Gmel. (American Planer-tree.) Leaves ovate-oblong, small, 1 to 1½ in. long, on short stems, sharp-pointed, serrate with equal teeth, smooth, green above and gray below, not oblique at base. Flowers minute, in small heads, appearing before the leaves. Fruit a scaly, roughened nut, ¼ in., raised on a stalk in the calyx; ripe in September. A small tree, 20 to 50 ft. high; wet banks, Kentucky and southward; hardy as far north as Philadelphia.

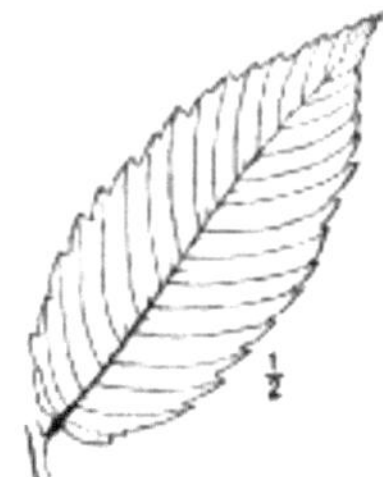

P. acuminàta.

2. **Plánera acuminàta.** (Kiaka Elm or Japan Planer-tree.) Leaves large, glossy, smooth, deeply notched, on red stems; young shoots also red. This is a larger, more hardy, and finer tree than the American Planer-tree, and should be more extensively cultivated.

The Caucasian Planer-tree (*Planera parvifolia*), with very small leaves, is also occasionally cultivated.

Genus **76. CÉLTIS.**

Trees or shrubs with alternate, simple, 2-ranked, oblique, serrate leaves. Flowers inconspicuous, greenish, axillary. Fruit berry-like, sweet, edible drupes, about the size of a currant, with one seed; color dark; ripe in autumn.

* Leaves usually sharply serrate   1.

* Leaves almost entire   2.

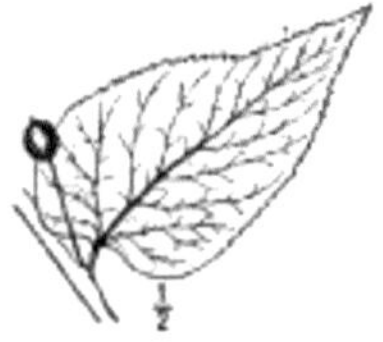

C. occidentàlis.

1. **Céltis occidentàlis**, L. (Sugarberry. Hackberry.) Leaves ovate, obliquely subcordate to truncate at base, long-acuminate, serrate (at least near the apex), rough above and hairy beneath. Fruit a single-seeded, ¼ in., globular drupe, solitary on a peduncle, 1 in. long, in the axils of the leaves; purple when ripe in autumn. [Pg 137]

Shrub (var. *pumila*) to large tree, 6 to 50 ft. high; throughout; rare north, abundant south. Sometimes cultivated. The branches are numerous, slender, horizontal, giving the tree a wide-spreading, dense top.

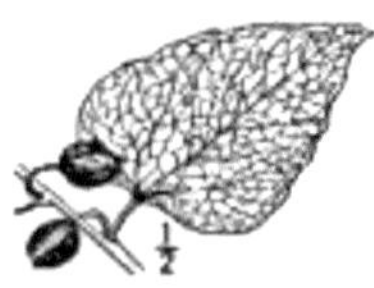

C. Mississippiénsis.

2. **Céltis Mississippiénsis**, Bosc. Leaves almost entire, with a very long, tapering point, a rounded and mostly oblique base, thin and smooth. Fruit smaller than that of the preceding species. A small tree with rough, warty bark. Illinois and southward.

Genus 77. MACLÙRA.

Trees or shrubs with milky juice and simple, alternate, entire, deciduous leaves, generally having a sharp spine by the side of the bud in the axils. Flowers inconspicuous; in summer. Fruit large, globular, orange-like in appearance.

M. aurantìaca.

Maclùra aurantìaca, Nutt. (Osage Orange. Bow-wood.) Leaves rather thick, ovate to ovate-oblong, almost entire, smooth and shining above, strong-veined and paler beneath, 4 in. long by 2 in. wide; spines simple, about 1 in. long. Fruit as large as an orange, golden-yellow when ripe. A medium-sized tree, 20 to 50 ft. high; native west of the Mississippi. Extensively cultivated for hedges, and also for ornament, throughout.

Genus **78. MÒRUS.**

Trees with milky juice and alternate, deciduous, exstipulate, broad, heart-shaped, usually rough leaves. Flowers inconspicuous; in spring. Fruit blackberry-like in shape and size; in summer.

* Leaves rough; fruit dark-colored   1.

* Leaves smooth and shining; fruit white to black   2.

[Pg 138]

M. rùbra.

1. **Mòrus rùbra**, L. (Red Mulberry.) Leaves broad, heart-shaped, 4 to 6 in. long, serrate, rough above and downy beneath, pointed; on the young shoots irregularly lobed. Fruit dark red, almost purple when ripe, cylindrical; not found on all the trees, as the flowers are somewhat diœcious; ripe in July. Wood yellow, heavy and durable. Usually a small tree, 15 to 60 ft. high; wild throughout, also cultivated.

M. álba.

2. **Mòrus álba**, L. (White Mulberry.) Leaves obliquely heart-ovate, pointed, serrate, smooth and shining; lobed on the younger growths; 2 to 7 in. long. Fruit whitish, oval to oblong; ripe in July. A small tree from China, planted for feeding silkworms, but now naturalized throughout.

Var. *multicaulis* has large leaves, and is considered better for silkworm food than the usual form. It is not very hardy, as it is frequently winter-killed in the latitude of New York City.

Var. *Downingii* (Downing's everbearing Mulberry) has large leaves and very large, dark red or black fruit, of excellent flavor, which does not ripen all at once as most Mulberries do.

Genus **79. BROUSSONÈTIA.**

Trees with milky juice and alternate, deciduous, stipulate, broad, very hairy leaves. Flowers diœcious. Fruit (only on a portion of the plants) similar to the common Mulberry.

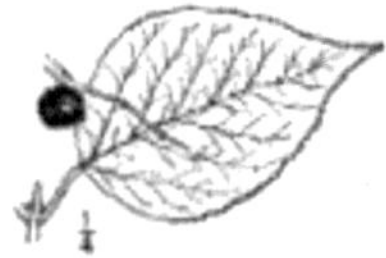

B. papyrífera.

Broussonètia papyrífera, L. (Paper-mulberry.) Leaves ovate to heart-shaped, variously lobed, deeply so on the young suckers, serrate, very rough above and quite soft-downy beneath; leaves on the old trees almost without lobes; bark tough and fibrous. Flowers in catkins, greenish; in spring. Fruit club-shaped, dark scarlet, sweet and insipid; ripe in August. Small cultivated tree, 10 to 35 ft. high, hardy north to New York; remarkable for the great variety in the forms of its leaves on the young trees. [Pg 139]

Order **XXXVII. PLATANÀCEÆ.**

(Plane-tree Family.)

A very small order, containing but one genus:

Genus **80. PLÁTANUS.**

Trees with alternate, simple, large, palmately lobed leaves. The base of the petiole is hollowed to cover the bud. Flowers inconspicuous; in early spring. Fruit a large, dry ball, hanging on a long peduncle, and remaining on the tree through the winter. Large tree with white bark separating into thin, brittle plates.

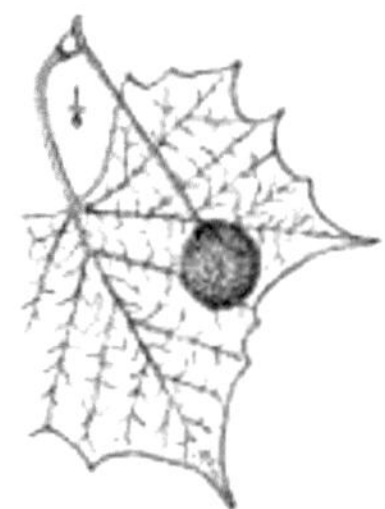

P. occidentàlis.

1. **Plátanus occidentàlis**, L. (American Sycamore. Buttonwood.) Leaves large (6 to 10 in. broad), roundish heart-shaped, angularly sinuate-lobed, the short lobes sharp-pointed, scurfy-downy till old. Fruit globular, solitary, 1 in. in diameter, hanging on long, 4-in. peduncles; remaining on the tree through the winter. A large, well-known tree, 80 to 100 ft. high; found on river-banks throughout; also cultivated. Wood brownish, coarse-grained; it cannot be split, and is very difficult to smooth. The marking of the grain on the quartered lumber is very beautiful.

P. orientàlis.

2. **Plátanus orientàlis**, L. (Oriental Plane.) Leaves more deeply cut, smaller, and sooner smooth than those of the American Sycamore. Fruit frequently clustered on the peduncles. This tree is similar to the American Sycamore, and in many ways better for cultivation. [Pg 140]

Order **XXXVIII. JUGLANDÀCEÆ.**

(Walnut Family.)

A small order of useful nut-and timber-trees.

Genus **81. JÙGLANS.**

Trees with alternate, odd-pinnate leaves, of 5 to 17 leaflets, with 2 to 4 axillary buds, the uppermost the largest. Flowers inconspicuous, the sterile ones in catkins. May. Fruit a large, bony, edible nut surrounded by a husk that has no regular dehiscence. The nut, as in the genus Carya, has a bony partition between the halves of the kernel.

* Leaflets 13 to 17, strongly serrate; husk of the fruit not separating from the very rough, bony nut; native. (**A.**)

    A. Upper axillary bud cylindrical, whitish with hairs; nut elongated   1.

    A. Upper axillary bud ovate, pointed; nut globular   2.

* Leaflets 5 to 9; husk of the fruit separating when dry from the smoothish, thin-shelled nut; cultivated   3.

J. cinèrea.

1. **Jùglans cinèrea**, L. (Butternut. White Walnut.) Leaflets 11 to 17, lanceolate, rounded at base, serrate with shallow teeth; downy, especially beneath; leafstalk sticky or gummy. Buds oblong, white-to-mentose. Fruit oblong, clammy, pointed. A thick-shelled nut, deeply sculptured and rough with ragged ridges; ripe in September. A widely spreading, flat-topped tree, 30 to 70 ft. high, with gray bark and much lighter-colored wood than that of the Juglans nigra. [Pg 141]

J. nìgra.

2. **Jùglans nìgra**, L. (Black Walnut.) Leaflets 13 to 21, lanceolate-ovate, taper-pointed, somewhat heart-shaped and oblique at base, smooth above and very slightly downy beneath. Fruit globular, roughly dotted; the thick-shelled nut very rough; ripe in October. A large handsome tree, 50 to 120 ft. high, with brown bark; more common west than east of the Alleghanies; often planted. Wood dark purplish-brown.

J. règia.

3. **Jùglans règia**, L. (Madeira Nut. English Walnut.) Leaflets 5 to 9, oval, smooth, obscurely serrate. Fruit oval, with a thin-shelled oval nut not nearly so rough as that of Juglans cinerea, or of Juglans nigra. When ripe the husk becomes very brittle and breaks open to let out the nut. Tree intermediate in size, 40 to 60 ft. high, hardy as far north as Boston in the East, but needs protection at St. Louis. It should be more extensively cultivated. Introduced from Persia.

Genus **82. CÁRYA.**

Hard-wooded trees with alternate, odd-pinnate leaves having straight-veined leaflets. The leaflets are opposite each other, and the terminal pair and end leaflet are usually much the largest. The sterile flowers are in hanging catkins, the fertile ones minute, forming a large, rounded, green-coated, dry drupe, with a roughened nut having a bony partition. The drupes hang on till frost, when they open more or less and usually allow the nut to drop out. Wood hard and tough. [Pg 142]

* Bark shaggy and scaly; kernel very good. (**A.**)

  A. Leaflets usually 5 (5 to 7)   1.

  A. Leaflets 7 to 9   2.

* Bark rough, deeply furrowed but not shaggy; kernel edible. (**B.**)

  B. Leaflets 7 to 9, usually 7   3.

  B. Leaflets 5 to 7, usually 5   4.

* Bark smooth; kernel bitter. (**C.**)

  C. Leaflets 5 to 7, usually 7, smooth   5.

  C. Leaflets 7 to 11, serrate with deep teeth   6.

* Bark smooth; nut thin-shelled; kernel sweet; leaflets 13 to 15   7.

C. álba.

1. **Cárya álba**, Nutt. (Shellbark or Shagbark Hickory.) Leaflets 5, the lower pair much smaller, all oblong-lanceolate, taper-pointed, finely serrate, downy beneath when young. Fruit globular, depressed at the top, splitting readily into 4 wholly separate valves. Nut white, sweet, compressed, 4-angled. Husk quite thin for the Hickories. Tree 70 to 90 ft. high, with very shaggy bark, even on quite small trees. Wild throughout, and cultivated.

C. sulcàta.

2. **Cárya sulcàta**, Nutt. (Big Shellbark. Kingnut.) Leaflets 7 to 9, obovate-acuminate, sharply serrate, the odd one attenuate at base and nearly sessile; downy beneath (more so than Carya alba). Fruit large, oval, 4-ribbed above the middle, with 4 intervening depressions. Husk very thick, entirely separating into 4 valves. Nut large, 1¼ to 2 in. long, dull-whitish, thick-shelled, usually strongly pointed at both ends. Kernel sweet and good. Tree 60 to 90 ft. high, with a shaggy bark of loose, narrow strips on large trees. Quite common west of the Alleghanies.

C. tomentòsa.

3. **Cárya tomentòsa**, Nutt. (Mockernut. White-heart Hickory.) Leaflets 7 to 9 (mostly 7), lance-obovate, pointed, obscurely [Pg 143] serrate or almost entire, the lower surface as well as the twigs and the catkins tomentose when young. Fruit globular or ovoid, usually with a very hard, thick husk slightly united at base. Nut somewhat hexagonal, with a very thick shell and well-flavored kernel. A tall, slender tree, 60 to 100 ft. high, with a rough deeply furrowed, but not shaggy bark. Common on dry hillsides throughout.

C. microcárpa.

4. **Cárya microcárpa**, Nutt. (Small Mockernut.) Leaflets about 5 (5 to 7), oblong-lanceolate, long-pointed, finely serrate, smooth, glandular beneath; buds small, ovate. Fruit small, subglobose, with a thin husk; nut not sharply angled, with a thin shell; edible. A large tree, 70 to 90 ft. high; New York, Pennsylvania, and westward.

C. porcìna.

5. **Cárya porcìna**, Nutt. (Pignut. Broom-hickory.) Leaflets 5 to 7 (usually 7), oblong-ovate, acuminate, serrate, smooth. Fruit pear-shaped to oval, somewhat rough, splitting regularly only about half-way. Nut large (1½ to 2 in. long), brownish, somewhat obcordate, with a thick, hard shell, and poor, bitter kernel. Tall tree, 70 to 80 ft. high, with dark-colored heart-wood, and rather smooth bark. Common on ridges.

C. amàra.

6. **Cárya amàra**, Nutt. (Bitternut. Swamp-hickory.) Leaflets 7 to 11, lan [Pg 144] ceolate to oblong-lanceolate, serrate with deep teeth. Fruit roundish-ovate, regularly separable only half-way, but friable at maturity. Nut small, white, subglobose, with a very thin shell and an extremely bitter kernel. Large tree with orange-yellow winter buds, and firm, not scaly, bark. Wild throughout, and sometimes cultivated.

C. olivæfórmis.

7. **Cárya olivæfórmis**, Nutt. (Pecan-nut.) Leaflets 13 to 15, ovate-lanceolate, serrate; lateral ones nearly sessile and decidedly curved. Fruit oblong, widest above the middle, with 4 distinct valves. Nut oblong, 1¼ in., nearer smooth than the other edible Hickory-nuts, the shell thin, but rather too hard to be broken by the fingers. The kernel is full, sweet, and good. A tall tree, 80 to 90 ft. high. Indiana and south; also cultivated, but not very successfully, as far north as New York City.

Order **XXXIX. CUPULÌFERÆ.** (Oak Family.)

This order contains more species of trees and shrubs in temperate regions than any other, except the Coniferæ. The genus Quercus (Oak) alone contains about 20 species of trees in the region covered by this work.

Genus **83. BÉTULA.**

Trees or shrubs with simple, alternate, mostly straight-veined, thin, usually serrate leaves. Flowers in catkins, opening in early spring, in most cases before the leaves. Fruit a leafy-scaled catkin or cone, hanging on till autumn. Twigs usually slender, the bark peeling off in thin, tough layers, and having peculiar horizontal marks. Many species have aromatic leaves and twigs. [Pg 145]

* Trunks with chalky white bark. (**A.**)

  A. Native. (**B.**)

    B. Small tree with leafstalks about ½ as long as the blades   1.

    B. Large tree; leafstalks about 1/3 as long as the blades   2.

  A. Cultivated; from Europe; many varieties   3.

190

* Bark not chalky white, usually dark. (**C.**)

  C. Leaves and bark very aromatic. (**D.**)

    D. Bark of trunk yellowish and splitting into filmy layers   5.

    D. Bark not splitting into filmy layers   4.

  C. Leaves not very aromatic; bark brownish and loose and shaggy on the main trunk; growing in or near the water   6.

B. populifòlia.

1. **Bétula populifòlia**, Ait. (American White or Gray Birch.) Leaves triangular, very taper-pointed, and usually truncate or nearly so at the broad base, irregularly twice-serrate; both sides smooth and shining, when young glutinous with resinous glands; leafstalks half as long as the blades and slender, so as to make the leaves tremulous, like those of the Aspen. Fruit brown, cylindrical, more or less pendulous on slender peduncles. A small (15 to 30 ft. high), slender tree with an ascending rather than an erect trunk. Bark chalky or grayish white, with triangular dusky spaces below the branches; recent shoots brown, closely covered with round dots.

B. papyrífera.

2. **Bétula papyrífera**, Marsh. (Paper or Canoe Birch.) Leaves 2 to 4 in. long, ovate, taper-pointed, heart-shaped, abrupt or sometimes wedge-shaped at the base, sharply and doubly serrate, smooth and green above, roughly reticulated, glandular-dotted and slightly hairy beneath; footstalk not over 1/3 the length of the blade. Fruit long-stalked and drooping. A large tree, 60 to 75 ft. high, with white bark splitting freely into very thin, tough layers. A variety, 5 to 10 ft. high (var. *minor*), occurs only in the White Mountains. Young shoots reddish or purplish olive-green deepening to a dark copper bronze. New England and westward, also cultivated. [Pg 146]

B. álba.

3. **Bétula álba**, L. (European White Birch.) Leaves ovate, acute, somewhat deltoid, unequally serrate, often deeply cut, nearly smooth; in var. *pubescens* covered with white hairs. Fruit brown, cylindric, drooping. A tree, 30 to 60 ft. high, with a chalky-white bark; from Europe, extensively cultivated in this country, under many names, which indicate the character of growth or foliage; among them may be mentioned *pendula* (weeping), *laciniata* (cut-leaved), *fastigiata* (pyramidal), *atropurpurea* (purple-leaved), and *pubescens* (hairy-leaved).

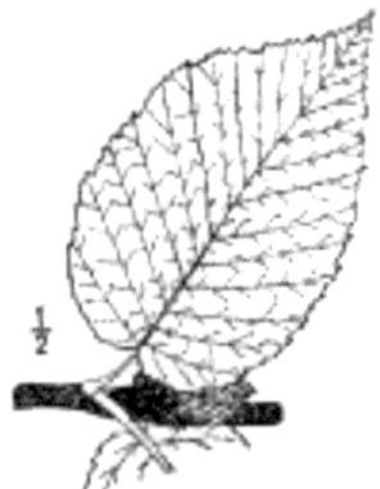

B. lénta.

4. **Bétula lénta**, L. (Sweet, Black or Cherry Birch.) Leaves and bark very sweet, aromatic. Leaves ovate or ovate-oblong, with more or less heart-shaped base, very acute apex, and doubly and finely serrate margin, bright shining green above, smooth beneath, except the veins, which are hairy. Fruit 1 to 1¼ in. long, cylindric, with spreading lobes to the scales. A rather large tree, 50 to 70 ft. high, with bark of trunk and twigs in appearance much like that of the garden Cherry, and not splitting into as thin layers as most of the Birches. Wood rose-colored, fine-grained. Moist woods, rather common throughout; also cultivated.

B. lùtea.

5. **Bétula lùtea**, Michx. f. (Yellow or Gray Birch.) A species so like the preceding (Betula lenta) as to be best described by stating the differences. Leaves and bark are much less aromatic. Leaves 3 to 5 in. long, not so often nor so plainly heart-shaped at base, usually narrowed; less bright green above, and more downy beneath; more coarsely serrate. Fruit not so long, and more ovate, with much larger and thinner scales, the lobes hardly spreading. A large tree, 50 to 90 ft. high, with yellowish or silvery-gray bark peeling off into very thin, filmy layers from the trunk. Wood whiter, and not so useful. Rich, moist woodlands, especially northward; also cultivated. [Pg 147]

B. nìgra.

6. **Bétula nìgra**, L. (River or Red Birch.) Leaves 2½ to 3½ in. long, rhombic-ovate, acute at both ends, distinctly doubly serrate, bright green above; glaucous beneath when young; on petioles only 1/6 their length. Twigs brown to cinnamon-color, and downy when young. A medium-sized tree, 30 to 50 ft. high, usually growing on the edges of streams, the old trunks having a very shaggy, loose, torn, reddish-brown bark. Wild in Massachusetts, south and west; often cultivated.

Genus **84. ÁLNUS.**

Shrubs or small trees with deciduous, alternate, simple, straight-veined leaves with large stipules that remain most of the season. Flowers in catkins. Fruit a small, scaly, open, woody cone, remaining on the plant throughout the year.

* Native species; growing in wet places. (**A.**)

    A. Leaves rounded at base; whitened beneath; found north of 41° N. Lat   1.

    A. Leaves acute or tapering at base; southward. (**B.**)

        B. Flowering in the spring   2.

        B. Flowering in the autumn   3.

* Cultivated species; from Europe; will grow in dry places   4, 5.

A. incàna.

1. **Álnus incàna**, Willd. (Speckled or Hoary Alder.) Leaves 3 to 5 in. long, broadly oval or ovate, rounded at base, sharply serrate, often coarsely toothed, whitened and mostly downy beneath; stipules lanceolate and soon falling. Fruit orbicular or nearly so. A shrub or small tree, 8 to 20 ft. high, with the bark of the trunk a polished reddish green; common along water-courses north of 41° N. Lat.; sometimes cultivated. [Pg 148]

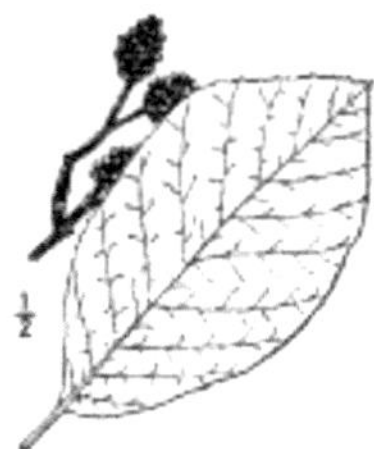

A. serrulàta.

2. **Álnus serrulàta**, Willd. (Smooth Alder.) Leaves 2 to 4½ in. long, thickish, obovate, acute at base, sharply and finely serrate, green both sides, smooth or often downy beneath; stipules yellowish green, oval, and falling after 2 or 3 leaves have expanded above them. Fruit ovate. Rather a shrub than a tree, 6 to 12 ft. high, common along streams south of 41° N. Lat. In the Southern States it sometimes forms a tree 30 ft. high.

A. marítima.

3. **Álnus marítima**, Muhl. (Seaside Alder.) Smooth; leaves oblong-ovate to obovate, with a tapering base, sharply serrulate; petiole slender; color bright green, somewhat rusty beneath. Flowering in the autumn. Fruiting catkin large, ¾ to 1 in. long, ½ in. thick, usually solitary, ovoid to oblong. A small tree, 15 to 25 ft. high. Southern Delaware and eastern Maryland, near the coast.

A. glutinòsa.

4. **Álnus glutinòsa**, L. (European Alder.) Leaves roundish, wedge-shaped, wavy-serrated, usually abrupt at tip, glutinous; sharply and deeply incised in some varieties. Fruit oval, ½ in. long. A medium-sized tree, 25 to 60 ft. high, of rapid growth, often cultivated under several names; the most important being vars. *laciniata* (cut-leaved), *quercifolia* (oak-leaved), and *rubrinervis* (red-leaved).

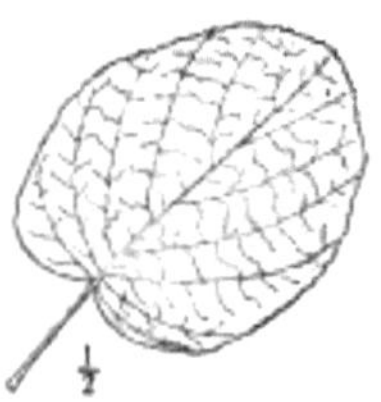

196

A. cordifòlia.

5. **Álnus cordifòlia**, Ten. (Heart-leaved Alder.) Leaves heart-shaped, dark green and shining. Flowers greenish-brown, blooming in March and April, before the leaves expand. A large and very handsome Alder, 15 to 20 ft. high, growing in much dryer soil than the Americàn species. Cultivated from southern Europe. Hardy after it gets a good start, but often winter-killed when young. [Pg 149]

Genus **85. CÓRYLUS.**

Low trees and large shrubs with simple, alternate, deciduous, doubly serrate, straight-veined leaves. Flowers insignificant, in catkins in early spring. Fruit an ovoid-oblong bony nut, inclosed in a thickish involucre of two leaves with a lacerated frilled border; ripe in autumn.

* Leafy bracts of fruit forming a bottle-shaped involucre   2.

* Leafy bracts not bottle-shaped. (**A.**)

    A. Involucre much longer than the nut   1.

    A. Involucre but little longer than the nut   3.

C. Americàna.

1. **Córylus Americàna**, Walt. (Wild Hazelnut.) Leaves roundish heart-shaped, pointed, doubly serrate; stipules broad at base, acute, and sometimes cut-toothed; twigs and shoots often hairy. Involucre of the fruit open to the globose nut, the two leaf-like bracts very much cut-toothed at the margin and thick and leathery at the base. Merely a shrub, 5 to 6 ft. high; quite common throughout.

C. rostràta.

2. **Córylus rostràta**, Ait. (Beaked Hazelnut.) Leaves but little or not at all heart-shaped; stipules linear-lanceolate. The involucre, extending beyond the nut in a bract like a bottle, is covered with stiff, short hairs. Shrub, 4 to 5 ft. high. Wild in the same region as Corylus Americana, but not so abundant.

C. Avellàna.

3. **Córylus Avellàna**, L. (European Hazel. Filbert.) Leaves roundish-cordate, pointed, doubly serrate, nearly sessile, with ovate-oblong, obtuse stipules; shoots bristly. Involucre of the fruit not much larger than the large nut (1 in.), and deeply cleft. A small tree or shrub, 6 to 12 ft. high, from Europe; several varieties in cultivation. [Pg 150]

Genus **86. ÓSTRYA.**

Slender trees with very hard wood, brownish, furrowed bark, and deciduous, alternate, simple, exstipulate, straight-veined leaves. Flowers inconspicuous, in catkins. Fruit hop-like in appearance, at the ends of side shoots of the season, hanging on through the autumn.

O. Virgínica.

1. **Óstrya Virgínica**, Willd. (Iron-wood. American Hop-hornbeam.) Leaves oblong-ovate, taper-pointed, very sharply doubly serrate, downy beneath, with 11 to 15 straight veins on each side of the midrib; buds acute. The hop-like fruit 2 to 3 times as long as wide; full grown and pendulous, 1 to 3 in. long, in August, when it adds greatly to the beauty of the tree. A small, rather slender tree, 30 to 50 ft. high, with the bark on old trees somewhat furrowed; wood white and very hard and heavy; common in rich woods, and occasionally cultivated.

O. vulgàris.

2. **Óstrya vulgàris**, Willd. (European Hop-hornbeam.) This species from Europe is much like the American one, but has longer, more slender, more pendulous fruit-clusters. Occasionally cultivated.

Genus **87. CARPÌNUS.**

Trees or tall shrubs with alternate, simple, straight-veined leaves, and smooth and close gray bark. Flowers in drooping catkins, the sterile flowers in dense cylindric ones, and the fertile flowers in a loose terminal one forming an elongated, leafy-bracted cluster with many, several-grooved, small nuts, hanging on the tree till late in the autumn. [Pg 151]

C. Caroliniàna.

1. **Carpìnus Caroliniàna**, Walt. (American Hornbeam. Blue or Water Beech.) Leaves ovate-oblong, pointed, sharply doubly serrate, soon nearly smooth. Fruit with the scales obliquely halberd-shaped and cut-toothed, ¾ in. long, nuts 1/8 in. long. A tree or tall shrub, 10 to 25 ft. high, with a peculiarly ridged trunk; the close, smooth gray bark and the leaves are much like those of the Beech. The wood is very hard and whitish. Common along streams; sometimes cultivated.

C. Bétulus.

2. **Carpìnus Bétulus**, L. (European Hornbeam.) This cultivated species is quite similar to the American, but can be distinguished by the scales of the fruit, which are wholly halberd-shaped, having the basal lobes nearly equal in size, as shown in the cut; while the American species has scales only half halberd-shaped.

Genus **88. QUÉRCUS.**

Large trees to shrubs, with simple, alternate, deciduous or ever-green, entire to deeply lobed leaves. The leaves are rather thick and woody, and remain on the tree either all winter or at least until nearly all other deciduous leaves have fallen. Flowers insignificant; the staminate ones in catkins; blooming in spring. Fruit an acorn, which in the White, Chestnut, and Live Oaks matures the same year the blossoms appear; while in the Red, Black, and Willow Oaks the acorns mature the second year. They remain on the tree until late in autumn. The Oaks, because of their large tap-roots, can be trans-planted only when small. Most of the species are in cultivation. The species are very closely related, and a number of them quite readily hybridize; this is especially true of those of a particular group, as the White Oaks, Black Oaks, etc. [Pg 152]

There is no attempt in the Key to characterize the hybrids, of which some are quite extensively distributed. *Quercus heterophylla*, Michx. (Bartram's Oak), supposed to be a hybrid between *Quercus Phellos* and *Quercus rubra*, is found quite frequently from Staten Island southward to North Carolina.

* Cultivated Oaks from the Old World; bark rough; leaves
more or less sinuated or lobed. (**A.**)

 A. Acorn cup not bristly   20.

 A. Acorn cup more or less bristly   21.

* Wild species, occasionally cultivated. (**B.**)

 B. Leaves entire or almost entire, or merely 3- (rarely 5-)
lobed at the enlarged summit. (**C.**)

  C. Ends about equal, petioles very short. (**D.**)

  D. Leaves small (2 to 4 in. long), evergreen, bark
smooth, black (Live-oaks)   10.

  D. Leaves not evergreen in the North, somewhat
awned when young, bark very smooth, black and
never cracked (Willow-oaks). (**E.**)

   E. Down on the under side quite persistent   18.

   E. Under side soon smooth   19.

  C. Widened near the tip, somewhat obovate and the
end usually 3-lobed; bark quite black, smooth or

furrowed, but never scaly (Black-oaks). (**F.**)

    F. Leaves acute at base   16.

    F. Leaves abrupt or cordate at base   17.

B. Leaves distinctly straight-veined, sinuate rather than lobed, the teeth generally rounded and never awned; bark white, rough and scaling (Chestnut-oaks). (**G.**)

    G. Lobes rounded   5, 6, 7.

    G. Lobes rather acute   8, 9.

B. Leaves coarsely lobed, the lobes usually rounded, never awned; bark white or whitish-brown, cracking and scaling off in thin laminæ (White Oaks). (**H.**)

    H. Leaves crowded at the ends of the branchlets   4.

    H. Leaves not crowded   1, 2, 3.

B. Leaves more or less lobed, the lobes and teeth acute and bristle-pointed; petiole slender; base rather abrupt; bark dark-colored, smooth or furrowed, but never scaly (Red Oaks). (**I.**)

    I. Leaves smooth both sides, at least when mature   11, 12, 13.

    I. Leaves soft-downy beneath   14, 15.

[Pg 153]

Q. álba.

1. **Quércus álba**, L. (American White Oak.) Leaves short-stemmed, acute at base, with 3 to 9 oblong, obtuse, usually entire, oblique lobes, very persistent, many remaining on the tree through the winter; pubescent when young, soon smooth, bright green above. Acorns in the axils of the leaves of the year, ovoid-oblong, 1 in., in a shallow, rough cup, often sweet and edible. A large tree, 60

to 80 ft. high, with stem often 6 ft. in diameter; wood light-colored, hard, tough and very useful. Common throughout.

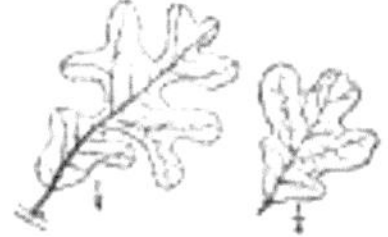

Q. stellàta

2. **Quércus stelláta**, Wang. (Post-oak. Rough or Box White Oak.) Leaves 4 to 6 in. long, sinuately cut into 5 to 7 roundish, divergent lobes, the upper ones much larger and often 1- to 3-notched, grayish-or yellowish-downy beneath, and pale and rough above. Acorn ovoid, about ½ in. long, one third to one half inclosed in a deep, saucer-shaped cup; in the axils of the leaves of the year. A medium-sized tree, 40 to 50 ft. high, with very hard, durable wood, resembling that of the White Oak. Massachusetts, south and west.

Q. macrocárpa.

3. **Quércus macrocárpa**, Michx. (Bur-oak. Mossy-cup.) Leaves obovate or oblong, lyrately pinnatifid or deeply sinuate-lobed or nearly parted, the lobes sparingly and obtusely toothed or entire. Acorn broadly ovoid, 1 in. or more long, one half to almost entirely inclosed in a thick and woody cup with usually a mossy fringed border formed of the upper awned scales; cup very variable in size, ¾ to 2 in. across. A handsome, middle-sized tree, 40 to 60 ft. high. Western New England to Wisconsin, and southwestward. [Pg 154]

Q. lyràta.

4. **Quércus lyràta**, Walt. (Swamp Post-oak.) Leaves crowded at the ends of the branchlets, very variable, obovate-oblong, more or less deeply 7- to 9-lobed, white-to-mentose beneath when young, becoming smoothish; the lobes triangular to oblong, acute or obtuse, entire or sparingly toothed. Acorn about ¾ in. long, nearly covered by the round, ovate, thin, rugged, scaly cup. A large tree with pale flaky bark. River-swamps in southern Indiana to Wisconsin, and southward.

Q. bícolor.

5. **Quércus bícolor**, Willd. (Swamp White Oak.) Leaves obovate or oblong-obovate, wedge-shaped at base, coarsely sinuate-crenate, and often rather pinnatifid than toothed, whitish, soft-downy beneath. Main primary veins 6 to 8 pairs. Acorns, nearly 1 in., oblong-ovoid, set in a shallow cup often mossy fringed at the margin, on a peduncle about as long as the acorn, much longer than the petioles of the leaves; in the axils of the leaves of the year. A large tree, 60 to 80 ft. high, stem 5 to 8 ft. in diameter. Most common in the Northern and Western States, in swamps, but found in moist soil in the mountains of the South.

Q. Michaùxii.

6. **Quércus Michaùxii**, Nutt. (Basket-oak or Cow-oak.) Leaves 5 to 6 in. long, oval to obovate, acute, obtuse, or even cordate at base, regularly but usually not deeply sinuate, rather rigid, usually very tomentose beneath. Acorn large, 1-1/3 in. long, sweet and edible; cup shallow and roughened with coarse, acute scales; no fringe. A large and valuable Oak with gray and flaky bark.

Q. Prìnus

7. **Quércus Prìnus**, L. (Chestnut-oak.) Leaves obovate or oblong, coarsely undulately toothed, with 10 to 16 pairs of straight, prominent ribs beneath; surface minutely downy beneath, and smooth above. Acorn ovoid, 1 in. long, covered nearly half-way [Pg 155] with a thick, mostly tuberculated cup; in the axils of the leaves of the year; kernel sweetish and edible. A middle-sized or small tree, with reddish, coarse-grained wood. Found throughout, but common only southward.

Q. Muhlenbérgii.

8. **Quércus Muhlenbérgii**, Engelm. (Yellow Chestnut-oak.) Leaves usually thin, 5 to 7 in. long, 1½ to 2 in. broad, oblong-lanceolate, rather sharply notched, mostly obtuse or roundish at base, sometimes broadly ovate or obovate, and two thirds as wide

as long. The leaves are usually more like those of the Chestnut than any other Oak; the primary veins very straight, impressed above, prominent beneath. Acorn 2/3 to ¾ in. long, inclosed in a thin, hemispherical cup with small, appressed scales. A middle-sized tree with flaky, pale, thin, ash-colored bark, and tough, very durable, yellowish or brownish wood. Western New England, westward and south.

Q. prinoìdes.

9. **Quércus prinoìdes**, Willd. (Dwarf Chestnut-oak.) Much like the last, but generally grows only 2 to 4 ft. high in the Eastern States. The leaves are more wavy-toothed, on shorter stems. It seems to be only a variety of Quercus Muhlenbergii, especially in the West, where it grows much taller and runs into that species.

Q. vìrens.

10. **Quércus vìrens**, Ait. (Live-oak.) Leaves thick, evergreen, 2 to 4 in. long, oblong, obtuse, and somewhat wrinkled; smooth and shining above, hairy beneath, the margin revolute, usually quite entire, rarely spiny-toothed. Acorns pedunculate, 1 to 3 in a cluster, oblong-ovate, with top-shaped nut. A mere shrub to a large tree, with yellowish wood of excellent grain and durability. Virginia and south. [Pg 156]

Q. rùbra.

11. **Quércus rùbra, L.** (Red Oak.) Leaves rather thin, smooth, oblong, moderately pinnatifid, sometimes deeply so, into 8 to 12 entire or sharply toothed lobes, turning dark red after frost. Acorn oblong-ovoid, 1 in. or less long, set in a shallow cup of fine scales, with a narrow raised border, ¾ to 1 in. in diameter; sessile or nearly so. A large tree, 60 to 90 ft. high, with reddish, very coarse-grained wood. Common throughout.

Q. coccínea.

12. **Quércus coccínea**, Wang. (Scarlet Oak.) Leaves, in the ordinary form on large trees, bright green, shining above, turning red in autumn, oval or oblong, deeply pinnatifid, the 6 to 8 lobes divergent, and sparingly cut-toothed, notches rounded. Acorn ½ to ¾ in. long, roundish, depressed, one half or a little more inclosed in a top-shaped, coarsely scaled cup; in the axils of the leaf-scars of the preceding year. A large handsome tree, 60 to 80 ft. high, with grayish bark not deeply furrowed, interior reddish; coarse-grained reddish wood. Moist or dry soil. Common.

Var. tinctória.

Var. *tinctoria*. (Quercitron. Yellow-barked or Black Oak.) Leaves, especially on young trees, often less deeply pinnatifid, sometimes barely sinuate. Foliage much like that of Quercus rubra. Acorn nearly round, ½ to 2/3 in. long, set in a rather deep, conspicuously scaly cup. Bark of trunk thicker, rougher, darker-colored and with the inner color orange. Rich and poor soil. Abundant east, but rare west.

Q. palústris.

13. **Quércus palústris**, Du Roi. (Swamp, Spanish, or Pin Oak.) Leaves oblong, deeply pinnatifid, with divergent, sharply toothed, bristle-tipped lobes and rounded notches, [Pg 157] and with both sides bright green. Acorn globular, hardly ½ in. long, cup shallow and saucer-shaped, almost sessile, in the axils of last year's leaf-scars. A handsome, medium-sized tree; wood reddish, coarse-grained. In low ground. Common throughout.

208

Q. falcàta.

14. **Quércus falcàta**, Michx. (Spanish Oak.) Leaves obtuse or roundish at base, 3- to 5-lobed above, the lobes prolonged, mostly narrow, and the end ones more or less scythe-shaped, bristle-tipped, entire or sparingly cut-toothed, soft-downy beneath. Foliage very variable. Acorn 1/3 to ½ in. long, globose, half inclosed in the hemispherical cup; nearly sessile. A tree, 30 to 70 ft. high, large and abundant in the South; bark thick and excellent for tanning; wood coarse-grained, dark brown or reddish. New Jersey, south and west.

Q. ilicifòlia.

15. **Quércus ilicifòlia**, Wang. (Bear or Black Scrub-oak.) Leaves obovate, wedge-shaped at base, angularly about 5-lobed (3 to 7), white-downy beneath, 2 to 4 in. long, thickish, with short, triangular bristle-tipped lobes. Acorn ovoid, globular, ½ in. long. A dwarfed, straggling bush, 3 to 10 ft. high. Sandy barrens and rocky hills. New England to Ohio, and south.

Q. aquática.

16. **Quércus aquática**, Walt. (Water-oak.) Leaves thick, sub-evergreen, obovate-wedge-shaped, smooth, tapering at the base, sometimes obscurely 3-lobed at the tip; on the seedlings and the young rapid-growing shoots often incised or sinuate-pinnatifid, and

then bristle-pointed. Acorn small, globular-ovoid, downy, in a saucer-shaped cup, very bitter; in the axils of leaf-scars of the previous year. A very variable tree, 30 to 40 ft. high, with smooth bark. Wet ground. Maryland, west and south. [Pg 158]

Q. nìgra.

17. **Quércus nìgra**, L. (Black Oak or Barren Oak.) Leaves large, 5 to 10 in. long, thick, wedge-shaped, broadly dilated above, and truncate or slightly 3-lobed at the end, bristle-awned, smooth above, rusty-downy beneath. Acorn oblong-ovate, ½ to ¾ in. long, in the axils of the leaves of the preceding year, one third or one half inclosed in the top-shaped, coarse-scaled cup. A small tree, 10 to 25 ft. high, with rough, very dark-colored bark. New York, south and west, in dry, sandy barrens.

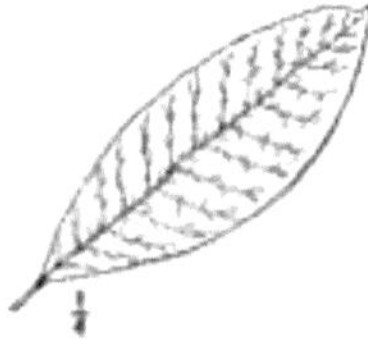

Q. imbricària.

18. **Quércus imbricària**, Michx. (Laurel-or Shingle-oak.) Leaves lanceolate-oblong, entire, tipped with an abrupt, sharp point, pale-downy beneath. Acorn globular, 5/8 in. long, cup with broad, whitish, close-pressed scales, covering about one third of the nut. A stout tree, 30 to 50 ft. high, found in barrens and open woodlands. Wood extensively used in the West for shingles. New Jersey to Wisconsin, and southward.

Q. Phéllos.

19. **Quércus Phéllos**, L. (Willow-oak.) Leaves 2 to 4 in. long, thick, linear-lanceolate, narrowed at both ends, entire or very nearly so, soon smooth, light green, bristle-tipped, willow-like, scurfy when young. Acorns about sessile, globular, small (½ in.), in a shallow saucer shaped cup; on the old wood. Tree 30 to 50 ft. high, with smooth, thick bark, and reddish, coarse-grained wood, of little value. Borders of swamps, New Jersey, south and west; also cultivated.

Q. Ròbur.

20. **Quércus Ròbur**, L. (English Oak.) Leaves on short footstalks, oblong, smooth, dilated upward, sinuately lobed, hardly pinnatifid. Acorns in the axils of the leaves of the year, ovate-oblong, over 1 in., about one third inclosed in the hemispherical cup; sessile in var. [Pg 159] *sessiliflora*; clustered and long-peduncled in var. *pedunculata*. Trees 50 to 100 ft. high, extensively cultivated; from Europe; the nursery catalogues name as many as a score or more varieties.

One var., *fastigiata* (Pyramidal Oak), is a peculiar upright tree like the Lombardy Poplar; var. *pendula* (Weeping Oak) has long, slender, drooping branches.

Q. Cérris.

21. **Quércus Cérris**, L. (Turkey Oak.) Leaves on very short stalks, oblong, deeply and unequally pinnatifid, hairy beneath; lobes lanceolate, acute, somewhat angular. Acorns in the axils of the leaves of the year, ovate, with a hemispherical, bristly or mossy cup. Several varieties of this species, from Europe, are cultivated in this country. They form tall, round-headed, symmetrical trees.

Genus **89. CASTÀNEA.**

Trees or shrubs with alternate, simple, straight-veined, elongated, pointed leaves. Sterile flowers in long, drooping, conspicuous catkins, blooming in June or July; the fertile ones rather inconspicuous, but forming prickly-coated burs which hang on till the frost, when they split open and let out the brown, horny-coated nuts. Wood light, coarse-grained.

* Large tree with burs having 1 to 3 nuts   1.

* Small tree with burs having 1 rounded nut   2.

[Pg 160]

C. satìva.

1. **Castànea satìva**, Mill. (Chestnut.) Leaves oblong-lanceolate, pointed, coarsely serrate, with usually awned teeth; smooth on both sides, 6 to 9 in. long, 1½ to 2¼ in. wide. Burs large, very prickly, inclosing 1 to 3 large, ovoid, brown nuts, ripe after frost, which

opens the bur into 4 valves. A common large tree, with light, coarse-grained wood, and bark having coarse longitudinal ridges on the old trees. Many varieties of this species are in cultivation, varying in the size and sweetness of the nuts, the size of the trees, and the size and the margins of the leaves, some of which are almost entire. The wild species is var. *Americana*.

C. pùmila.

2. **Castànea pùmila**, Mill. (Chinquapin.) Leaves lance-oblong, strongly straight-veined, coarsely serrate, usually with awned tips; whitish-downy beneath, 3 to 5 in. long, 1¼ to 2 in. wide. Bur small, prickly, with a single small, rounded, sweet, chestnut-colored nut. A handsome small tree, or in the wild state usually a shrub, 6 to 40 ft. high. Central New Jersey, southern Ohio and southward, and cultivated successfully as far north as New York City.

Genus **90. FÀGUS.**

Trees with alternate, strongly straight-veined, almost entire to deeply pinnatifid leaves. Flowers inconspicuous, appearing with the leaves. Fruit a prickly bur, inclosing 2 triangular, sharp-ridged nuts, the bur hanging on the trees during the greater part of the winter. Leaf-buds very elongated, slender, sharp-pointed.

* The straight veins all ending in the teeth; native   1.

* Margin varying from entire to deeply pinnatifid, the straight veins occasionally ending in the notches   2.

[Pg 161]

F. ferrugínea.

1. **Fàgus ferrugínea**, Ait. (American Beech.) Leaves thin, oblong-ovate, taper-pointed, distinctly and often coarsely toothed; petioles and midrib ciliate with soft silky hairs when young, soon almost naked. The very straight veins run into the teeth. Prickles of the fruit mostly recurved or spreading. Large tree, 60 to 100 ft. high, with grayish-white, very smooth bark, and firm, light-colored, close-grained wood. Wild throughout, and frequently cultivated.

F. sylvática.

2. **Fàgus sylvática**, L. (European Beech.) Leaves often similar to those of the American Beech, but usually shorter and broader; the border, often nearly entire, is wavy in some varieties, and in others deeply pinnatifid. The bark in most varieties is darker than in the American. This Beech, with its numerous varieties, is the one usually cultivated. Among the most useful varieties are *atropurpurea* (Purple Beech), with the darkest foliage of any deciduous tree, and almost entire-margined leaves; *laciniata* (Cut-leaved Beech), with very deeply cut leaves; and *argentea variegata* (Silver Variegated Beech), having in the spring quite distinctly variegated leaves.

Order **XL. SALICÀCEÆ.** (Willow Family.)

A small order of soft-wooded trees and shrubs, abundantly distributed in the northern temperate and frigid zones.

Genus **91. SÀLIX.**

Soft-wooded trees or shrubs growing in damp places, with alternate, usually quite elongated, pointed, deciduous [Pg 162] leaves, without lobes. Stipules often large, leaf-like, and more or less persistent through the summer; sometimes scale-like and dropping early. The stipules are always free from the leafstalk and attached to the twig at small spots just below the leafstalk. Even if the stipules have dropped off, the small scars remain. Flowers staminate and pistillate on separate trees (diœcious), in elongated catkins in early spring. Fruit consists of catkins of small pods with numerous seeds having silky down at one end. The seeds usually drop early. Among the Willows there are so many hybrids and peculiar varieties as to render their study difficult, and their classification, in some cases, impossible. The following Key will probably enable the student to determine most specimens. No attempt has been made to include all the cultivated forms.

* Spray decidedly weeping   5.

* Spray not decidedly weeping. (**A.**)

  A. Rather small Willows, 10 to 30 ft. high, with broad leaves, usually not over twice as long as wide; cultivated. (**B.**)

    B. Leaves glossy dark green on the upper side, taper-pointed   7.

    B. Leaves with white cottony hairs beneath   10.

    B. Leaves rough-veiny beneath   13.

  A. Rather large Willows, 12 to 80 ft. high, with the bark of the trunk very rough; leaves more elongated. (**C.**)

    C. Petioles of the leaves not glandular; tree 10 to 40 ft. high. (**D.**)

      D. Leaves green on both sides when mature   1.

      D. Leaves glaucous beneath   2.

    C. Petioles of the leaves usually glandular; tree 50 to 80 ft. high. (**E.**)

E. Young leaves green above and glaucous beneath 3.

E. Young leaves ashy gray or silvery white on both sides 4.

A. Small trees or almost shrubs, under 18 ft. high; bark of trunk rather smooth. (**F.**)

F. Leaves ovate rather than lanceolate, sometimes truncate or even cordate at base. (**G.**)

G. Leaves quite broad, shining on both sides. (**H.**)

H. Leaves bright green; twigs polished green 6.

H. Leaves very dark green, strongly fragrant when bruised 7. [Pg 163]

G. Leaves pale-downy beneath, often cordate at base 8.

F. Leaves usually wider near the acute or acuminate tip, glaucous beneath. (**I.**)

I. Branches very twiggy; leaves often opposite; twigs olive-color or reddish 9.

I. Branches not very twiggy; leaves all alternate 11, 12.

F. Leaves very long and slender, almost linear 14.

S. nìgra

1. **Sàlix nìgra**, Marsh. (Black Willow.) Leaves narrowly lanceolate, tapering at the ends, serrate, smooth except on the petiole and midrib, green on both sides; stipules small (large in var. *falcata*), dentate, dropping early. Branches very brittle at base. A small tree, 15 to 35 ft. high, with rough black bark. Common along streams, southward, but rare in the northern range of States.

S. amygdaloìdes.

2. **Sàlix amygdaloìdes**, Anderson. (Western Black Willow.) Leaves 2 to 4 in. long, lanceolate or ovate-lanceolate, attenuate-cuspidate, pale or glaucous beneath, with long slender petioles; stipules minute and soon falling. A small tree, 10 to 40 ft. high, from central New York westward. It is the common Black Willow of the streams of Ohio to Missouri.

S. frágilis.

3. **Sàlix frágilis**, L. (Brittle Willow. Crack-willow.) Leaves lanceolate, taper-pointed, smooth, glaucous beneath (slightly silky when young), serrate throughout; stipules half heart-shaped, usually large. Branches smooth and polished, very brittle at base. A tall (50 to 80 ft. high) handsome Willow, with a bushy head and salmon-colored wood; cultivated from Europe for basket-work, and extensively naturalized. Many varieties, hybrids between this species and the next, are very common. Among them may be mentioned the following: [Pg 164]

Var. *decipiens*, with dark-brown buds; var. *Russelliana*, with more slender, brighter, and more sharply serrate leaves, the annual shoots silky-downy toward autumn; var. *viridis*, with tough, pendulous branchlets, and firmer, bright green leaves.

S. álba.

4. **Sàlix álba**, L. (White Willow.) Leaves lanceolate or elliptical-lanceolate, pointed, serrate, covered more or less with white silky hairs, especially beneath; var. *cærulea* has nearly smooth leaves, at maturity of a bluish tint; stipules small and quite early deciduous. Catkins of flowers long and loose, on a peduncle; stamens usually 2; stigmas nearly sessile, thick, and recurved. May, June. A quite large tree, 50 to 80 ft. high, with thick, rough bark, usually having yellow twigs (var. *vitellina*); introduced from Europe and now quite common throughout. Branches very brittle at base.

S. Babylónica.

5. **Sàlix Babylónica**, Tourn. (Weeping Willow.) Leaves linear-lanceolate, acuminate, finely serrate, smooth, glaucous beneath; stipules small, roundish, oblique, acuminate; branches pendulous.

A large, gracefully drooping tree, so extensively cultivated for ornament as to seem native; from Europe.

Var. *annularis* (Ring-leaved Willow. Curled Willow) has the leaves coiled round into rings and spirals.

S. lùcida.

6. **Sàlix lùcida**, Mühl. (Shining Or American Bay Willow.) Leaves thickish, ovate-lanceolate, with a rounded base, a very long acuminate point, and a glandular petiole; when mature, smooth and shining on both sides. Twigs rather stout, polished, and dark green. Bark of trunk smooth. Fruiting catkins quite persistent. A beautiful small tree or shrub, 6 to 15 ft. high, of bushy form. New Jersey, north and westward. [Pg 165]

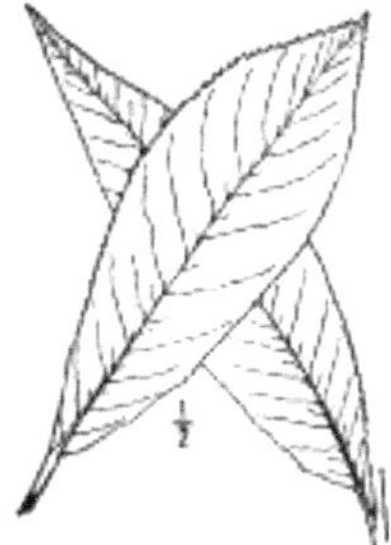

S. pentándra.

7. **Sàlix pentándra, L.** (Laurel-leaved or Bay Willow.) Leaves ovate, taper-pointed, crenate, glandular, smooth, glossy, bright deep green on both sides, strongly fragrant when bruised. Catkins large, fragrant, golden-yellow, with 4 to 12 (commonly 5) stamens to each flower. June, after the leaves are expanded. A small handsome tree, 15 to 20 ft. high, from Europe, which should be more extensively cultivated in damp soils, as its form, flowers, and foliage are all beautiful.

S. cordàta. Var. rufescens.

8. **Sàlix cordàta, Mühl.** (Heart-leaved Willow.) Leaves lanceolate or ovate-lanceolate, heart-shaped, truncate or sometimes acute at base, taper-pointed, sharply serrate, smooth above, pale-downy beneath; stipules often large, kidney-shaped, and toothed, sometimes small and entire. Catkins appearing with or before the leaves along the sides of the stem; stamens 2; scales dark or black, hairy, persistent. Shrub or small tree, 8 to 20 ft. high, very common in low and wet places. Many named varieties are found.

Var. *rigida* has large, thick, coarse-toothed leaves; vars. *myricoides* and *angustata* have narrower, finely serrate leaves, almost or fully acute at base.

S. purpùrea.

9. **Sàlix purpùrea, L.** (Purple Willow.) Leaves lanceolate, pointed, partly opposite, minutely serrate, smooth. Twigs olive-color [Pg 166] or reddish. Catkins cylindric, with leafy bracts at base, and apparently 1 stamen to each flower (the filaments are united). A shrub or small tree, 3 to 12 ft. high; from Europe. In low ground; often cultivated for the twigs, which are used in basket-making.

S. càprea.

10. **Sàlix càprea, L.** (Goat-willow.) Leaves large, roundish, ovate, pointed, serrate, wavy, deep green above, pale and downy with soft, white-cottony hairs beneath; stipules somewhat crescent-shaped. Catkins large, oval, numerous, almost sessile, blooming much before the leaves appear, and of a showy yellow color. A moderate-sized tree, 15 to 30 ft. high, with spreading, brown or purplish branches. Frequent in cultivation; from Europe; growing well in dry places. The Goat-willow is the one generally used for the stock of the artificial umbrella-formed "Kilmarnock Willow." The growth of shoots from these stocks is rendering the Goat-willow quite common.

S. rostràta.

11. **Sàlix rostràta, Richards.** (Beaked Willow.) Leaves oblong to obovate-lanceolate, acute, usually obscurely toothed, sometimes crenate or serrate, downy above, prominently veined, soft-hairy and somewhat glaucous beneath. Twigs downy. Catkins appearing with the leaves. Fruit-capsules tapering to a long slender beak, pedicels long and slender. A small, tree-shaped shrub, 4 to 15 ft. high, common in both moist and dry ground. New England, west and north.

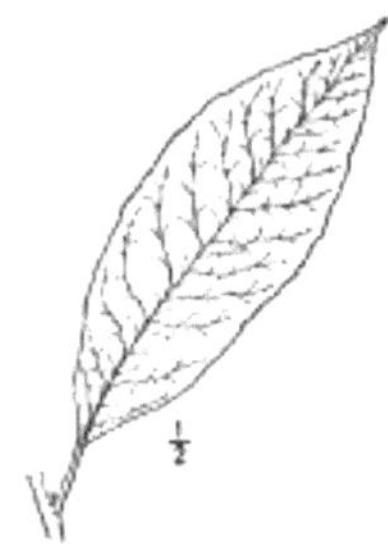

S. díscolor.

12. **Sàlix díscolor, Mühl.** (Glaucous or Bog Willow.) Leaves lanceolate or ovate-lanceolate, acute, remotely serrate at the base, finely

serrate along the middle, and almost entire near the tip; smooth and bright green above, soon smooth and somewhat glaucous beneath; stipules, on the vigorous shoots, equaling the petiole, more frequently small and inconspicuous. Catkins sessile, 1 in. long, appearing before the leaves in the spring; scales dark red or brown, becoming black, covered with long glossy hairs. Fruit in catkins, 2½ in. [Pg 167] long, the capsules very hairy, with short but distinct style. A very variable species, common in low meadows and on river-banks; usually a shrub, but occasionally 15 ft. high.

S. cinèrea.

13. **Sàlix cinèrea, L.** (Gray or Ash-colored Willow.) Leaves obovate-lanceolate, entire to serrate; glaucous-downy and reticulated with veins beneath; stipules half heart-shaped, serrate. Flowers yellow; ovary silky, on a stalk half as long as the bracts. A shrub to middle-sized tree, 10 to 30 ft. high, with an erect trunk; occasionally cultivated; from Europe.

S. longifòlia.

14. **Sàlix longifòlia,** Mühl. (Long-leaved Willow.) Leaves linear-lanceolate, very long, tapering at each end, nearly sessile, remotely notched with projecting teeth, clothed with gray hairs when young; stipules small, lanceolate, toothed. Branches brittle at base. A shrub

222

or small tree, 2 to 20 ft. high, common, especially westward, along river-banks.

Genus **92. PÓPULUS.**

Trees with alternate, deciduous, broad-based leaves. Flowers in long and drooping catkins, appearing before the leaves are expanded in the spring. Fruit small, dry pods in catkins, having seeds, coated with cottony down, which early in the season escape and float in the wind. On this account the trees are called Cottonwoods in the West. Trees with light-colored, rather soft wood.

* Leaves always white-hairy underneath; more or less deeply lobed; buds not gummy   1.

* Leaves smooth beneath, at least when old. (**A.**)

  A. Leafstalk decidedly flattened laterally. (**B.**)

    B. Buds not covered with sticky gum. (**C.**) [Pg 168]

      C. Leaves roundish heart-shaped; bark on trunk greenish-white,   2.

      C. Leaves large, ovate, with large, irregular, sinuate teeth,   3.

    B. Buds covered with aromatic, glutinous resin. (**D.**)

      D. Tree tall, spire-shaped,   5.

      D. Not very spire-shaped; young twigs sharply angled or winged, leaves 6 to 10 in. long, broadly deltoid, serrate with incurved teeth,   6.

      D. Not spire-shaped; young twigs not angular,   7.

  A. Leafstalk not decidedly flattened; leaf-margin crenate. (**E.**)

  E. Buds not glutinous; leaves white-woolly beneath when young,   4.

  E. Buds very glutinous; leaves large, shining green on both sides,   8.

P. álba.

1. **Pópulus álba**, L. (White Poplar or Abele Tree.) Leaves roundish, slightly heart-shaped, wavy toothed or lobed, soon green above, very white-cottony beneath even when old; buds without the sticky coating common in the genus. Branches very white with down when young. Root creeping and producing numerous suckers. A large tree, 50 to 80 ft. high, of rapid growth, often cultivated; from Europe. Leaves and branches very variable, forming several named varieties in the catalogues of the nurseries.

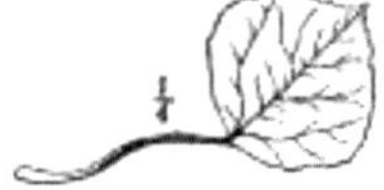

P. tremuloìdes.

2. **Pópulus tremuloìdes**, Michx. (Quaking-asp. American Aspen.) Leaves roundish heart-shaped, with a short sharp point, and small, quite regular teeth; downy when young, but soon smooth on both sides; margins downy. Leafstalk long, slender, compressed, causing the leaves to tremble continually in the slightest breeze. Leaf with 2 glands at the base on the upper surface; buds varnished. A medium-sized tree, 30 to 60 ft. high; bark greenish-white outside, yellow within, quite brittle. Common both in forests and in cultivation.

P. grandidentàta.

3. **Pópulus grandidentàta**, Michx. (Large-toothed Aspen.) Leaves large, 3 to 5 in. long, roundish-ovate, with large, irregular, sinu [Pg 169] ate teeth; and when young densely covered with white, silky wool, but soon becoming smooth on both sides; leaf, when young, reddish-yellow; petiole compressed. A large tree, 60 to 80 ft. high, with rather smoothish gray bark. Woods; common northward, rare southward, except in the Alleghanies. Wood soft and extensively used for paper-making.

P. heterophýlla.

4. **Pópulus heterophýlla**, L. (Downy-leaved Poplar.) Leaves heart-shaped or roundish-ovate with small, obtuse, incurved teeth; white-woolly when young, but soon becoming smooth on both sides except on the veins beneath. Leafstalk slightly compressed. Shoots round, tomentose. Buds not glutinous. A large tree, 70 to 80 ft. high, not very common; found from western New England to Illinois, and southward.

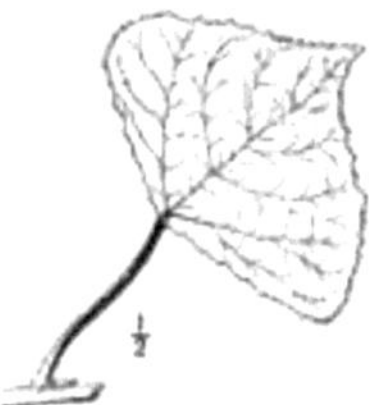

P. dilatàta.

5. **Pópulus dilatàta**, L. (Lombardy Poplar.) Leaves deltoid, wider than long, crenulated all round, both sides smooth from the first; leafstalk compressed; buds glutinous. A tall tree, 80 to 120 ft. high; spire-like, of rapid growth, with all the branches erect; the trunk twisted and deeply furrowed. Frequently planted a century ago, but now quite rare in the eastern United States. From Europe. It is thought to be a variety of Populus nigra (No. 7).

P. monilífera.

6. **Pópulus monilífera**, Ait. (Cottonwood. Carolina Poplar. Necklace-poplar.) Leaves large, broadly heart-shaped or deltoid, serrate with cartilaginous, incurved, slightly hairy teeth. The rapid-growing young twigs very angular and bearing very large (6 to 9 in. long) leaves. A very large (80 to 100 ft. high) tree, common in the Mississippi valley, but found in western New England and often planted. [Pg 170]

P. nìgra.

7. **Pópulus nìgra**, L. (Black Poplar.) Leaves rather large, deltoid, pointed, serrate with glandular teeth, smooth on both sides even when young. Leafstalk somewhat compressed. Buds very sticky. A very variable, large (50 to 80 ft. high), rapidly growing tree with spreading branches. Occasionally planted. From Europe.

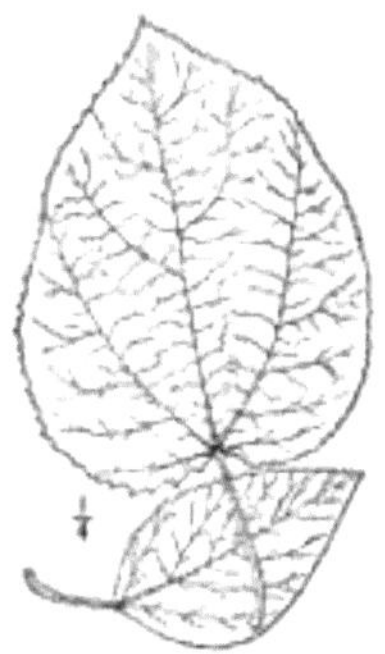

P. balsamífera.

8. **Pópulus balsamífera**, L. (Balsam-poplar. Tacamahac. Balm of Gilead.) Leaves very large, ovate, gradually acuminate, sometimes heart-shaped, finely serrate, smooth, bright green and shining on both sides; leafstalk nearly round; leaves in spring rich yellow. Branches ridged below the leaves; buds large and covered with very fragrant resin. A medium-sized tree, 40 to 70 ft. high, pyramidal in form. Wild in the North and often cultivated.

Var. *candicans*, or Balm of Gilead, has larger and more or less heart-shaped leaves (the larger figure in the cut).

CLASS II. GYMNOSPÉRMÆ.

Plants in which the pistil is represented by an open scale instead of a body with a closed ovary, as in Class I.

Order **XLI. CONÍFERÆ.** (Pine Family.)

As far as the number of species is concerned, this is the largest order of trees and shrubs of temperate and cold-temperate regions. The order is of the greatest importance, both on account of the valuable timber it furnishes and for its resinous secretions, turpentine and resin. [Pg 171]

Genus **93. PÌNUS.** (The Pines.)

Leaves needle-shaped, 1 to 15 in. long, almost cylindric, 2, 3, or 5 together in clusters, with a sheath, more or less persistent, at the base. Flowers monœcious, both staminate and pistillate in catkins, usually insignificant and unnoticeable. In spring. Fruit a cone, persistent and formed of more or less woody, overlapping scales.

* Leaves usually 5 together in bundles. (**A.**)

  A. Leaves 6 in. or more long, glaucous green and very pendulous — 1.

  A. Leaves under 4 in. long. (**B.**)

    B. Cones over 10 in. long, on stalks 3 in. long, pendulous when ripe — 2.

    B. Cones 4 to 10 in. long. (**C.**)

      C. Scales of cones thin, unarmed — 3, 4.

      C. Scales of cones thick and woody, obtuse, 1 in. broad — 5.

    B. Cones under 4 in. long; scales slightly hooked but pointless — 6.

* Leaves usually in threes, rarely in twos; scales of cones with spines or prickles. (**D.**)

  D. Scales of cones with short, rigid, straight spines; leaves 6 to 10 in. long — 7.

  D. Scales with sharp, bent prickles. (**E.**)

    E. Leaves over 5 in. long, sometimes 15 in. long — 8, 9.

    E. Leaves 3 to 5 in. long, rigid and flattened, from short sheaths, — 10.

* Leaves usually in twos; cones rarely over 3 in. long. (**F.**)

  F. Leaves over 3 in. long. (**G.**)

    G. Cone-scales with dull spines — 11.

    G. With small or minute, persistent prickles — 12, 13, 14.

    G. With no prickles, or small ones, early deciduous — 15, 16.

  F. Leaves 3 in. or less long. (**H.**)

    H. Cone-scales with straight or slightly curved, rigid spines — 17.

    H. Cone-scales with stout, recurved spines — 18, 19.

    H. Cone-scales with small prickles which are early deciduous — 20.

    H. Cone-scales without spines or prickles — 21, 22.

P. excélsa.

1. **Pìnus excélsa**, Wallich. (Bhotan Pine.) Leaves in fives, from short, fugacious, overlapping, membranaceous sheaths, 6 to 7 in. long, very slender, of a glaucous-green color, and very pendulous. Cones 6 to 9 in. long, and 2 in. in diameter, drooping and clustered, with broad, thick, wedge-shaped scales. A large beautiful tree from southern Asia, much subject to blight when planted in this country. Owing to its peculiar drooping branches it has been called the Weeping Fir.

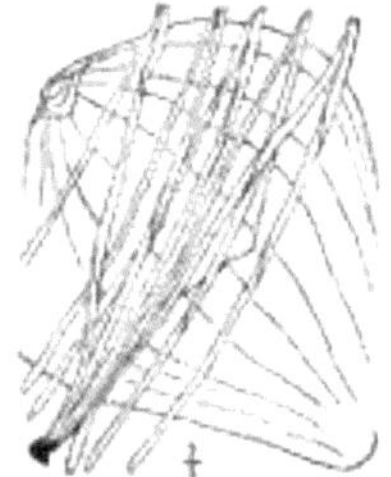

P. Lambertiàna.

2. **Pìnus Lambertiàna**, Douglas. (Lambert's or Sugar Pine.) Leaves in fives, 3 to 4 in. long, from short, deciduous sheaths. Cones 12 to 18 in. long and 3 to 4 in. in diameter, gradually tapering to a point, on stalks 3 in. long, brown and pendulous when ripe, without resin; seeds large, oval, nearly 1 in. long, edible. A very large tree (100 to 300 ft. high in California and northward), and seemingly hardy and well worth cultivation in the East. Wood white and soft like that of the White Pine.

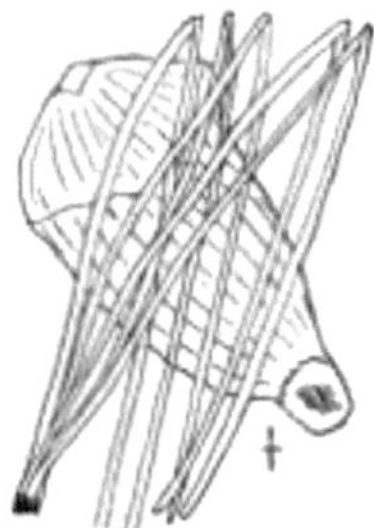

P. Stróbus.

3. **Pìnus Stróbus**, L. (White Pine. Weymouth Pine.) Leaves in fives, 3 to 4 in. long, from a loose, deciduous sheath; slender, soft, and whitish on the under side. Cones 4 to 6 in. long, cylindric, usually curved, with smooth, thin, unarmed scales. Tall (100 to 150 ft. high), very useful tree, of white, soft wood nearly free from resin and more extensively used for lumber than any other American tree. Has been common throughout, but is getting scarce on account of its consumption for lumber. [Pg 173]

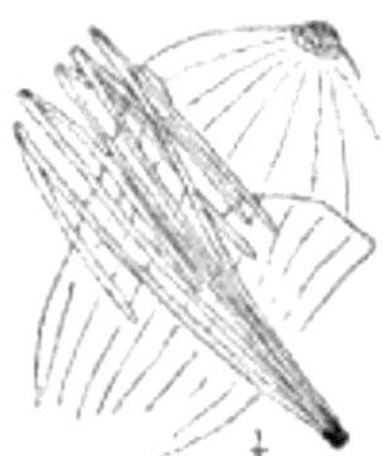

P. montícola.

4. **Pìnus montícola**, Dougl. (Mountain-pine.) Leaves in fives, 3 to 4 in. long, from short, overlapping, very deciduous sheaths; smooth, glaucous green. Cones 7 in. long and 1¾ in. in diameter, cylindric, smooth, obtuse, short-peduncled, resinous, with loosely overlapping, pointless scales. A large tree, 60 to 80 ft. high, resembling the White Pine, and often considered a variety of it, but the foliage is denser; Pacific coast.

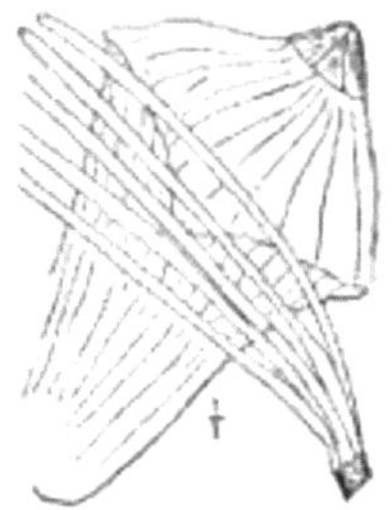

P. fléxilis.

5. **Pìnus fléxilis**, James. (Western White Pine.) Leaves 2 to 3 in. long, rigid, entire, acute, densely crowded, sharp-pointed, of a rich dark green color, 5 together in lanceolate, deciduous sheaths. Cones 4 to 6 in. long and half as wide, subcylindric, tapering to the end, semipendulous, clustered. Scales thick, woody, obtuse, loose, 1¼ in. broad, yellowish brown. Seeds rather large, with rigid margins instead of wings. A handsome hardy tree from the Pacific Highlands, occasionally cultivated. It resembles the eastern White Pine, but is more compact and of a darker color.

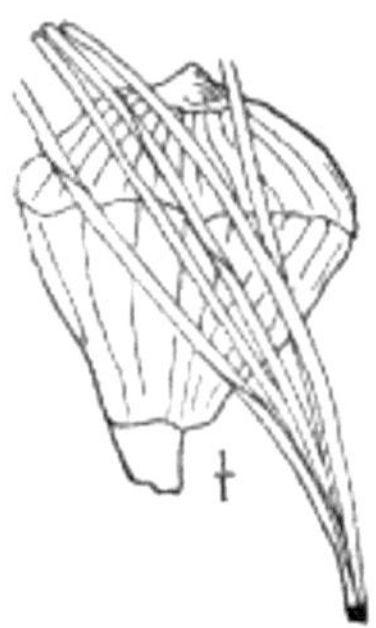

P. Cémbra.

6. **Pìnus Cémbra**, L. (Cembra Pine. Swiss Stone-pine.) Leaves 3 to 4 in. long, from a medium-sized deciduous sheath; triangular, rigid, slender, straight, crowded, dark green with a glaucous surface; 5 together. Cones 2½ in. by 2 in., ovate, erect, with obtuse, slightly hooked, but pointless scales. Seeds as large as peas and destitute of wings. A slow-growing, cultivated tree, 40 to 80 ft. high. Forms a regular cone; branches to the ground; Europe; hardy throughout. [Pg 174]

P. Tǽda.

7. **Pìnus Tǽda**, L. (Loblolly or Old-field Pine.) Leaves in twos and threes, 6 to 10 in. long, with elongated, close sheaths; slender and of a light green color. Cones in pairs or solitary, lateral, 3 to 4 in. long, oblong, conical; the scales having short, rigid, straight spines. A large tree, 50 to 130 ft. high, wild from Delaware, south and west, in swamps and old fields.

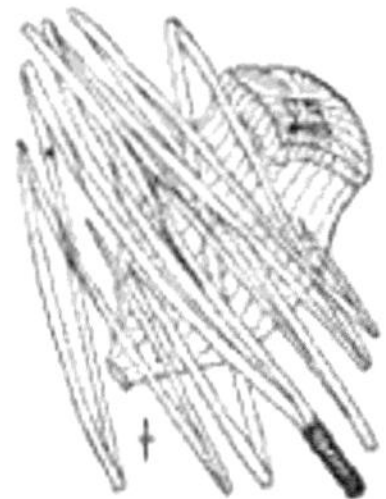

P. ponderòsa.

8. **Pìnus ponderòsa**, Dougl. (Western Yellow or Heavy-wooded Pine.) Leaves in threes, 5 to 10 in. long, from short sheaths; broad, coarse, twisted, flexible, of a deep green color; branchlets thick, reddish brown. Cones 3 to 4 in. long, ovate, reflexed, clustered on short stems. Scales long, flattened, with small, sharp, recurved prickles. A large Pacific coast species, 100 to 300 ft. high, with rather coarse-grained, hard and heavy, whitish wood, and thick, deeply furrowed bark; beginning to be cultivated east.

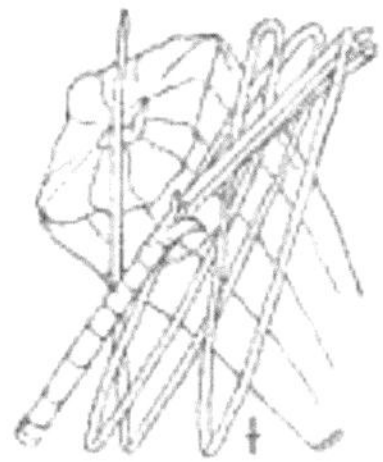

P. paltústris.

9. **Pìnus palústris**, Mill. (Long-leaved or Southern Yellow Pine.) Leaves 3 together in bundles, 10 to 15 in. long, from a long, lacerated, light-colored sheath, of a bright green color, and crowded in dense clusters at the ends of the branches. Cones 6 to 10 in. long, usually cylindric, of a beautiful brown color, with thick scales, armed with very small, slightly recurved prickles. A rather tall pine, 75 ft. high, wild in the Southern States, and cultivated as far north as New Jersey, in sheltered situations.

P. rígida.

10. **Pìnus rígida**, Mill. (Pitch-pine.) Leaves in threes, 3 to 5 in. long, from short sheaths; rigid and flattened. Cones ovate, 1 in. [Pg 175] to nearly 4 in. long, sometimes in clusters; scales with a short, recurved prickle. A medium-sized tree, 40 to 70 ft. high, with hard, coarse-grained, very resinous wood; found east of the Alleghanies throughout; more abundant in swamps.

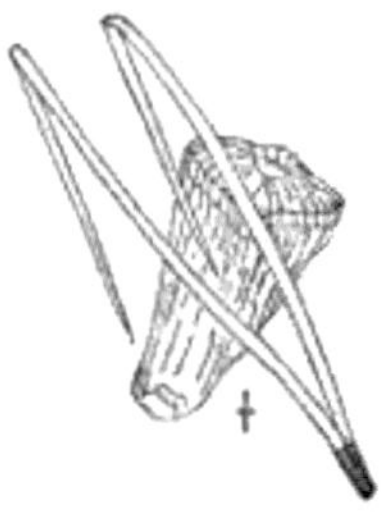

P. Austrìaca.

11. **Pìnus Austrìaca**, Höss. (Austrian or Black Pine.) Leaves long, 3 to 5 in., rigid, slender, incurved, sharply mucronate, of a dark green color; from short sheaths; 2 together. Cones 2½ to 3 in. long, regularly conical, slightly recurved, of a light brown color; scales smooth, shining, with a dull spine in the center. A large cultivated tree, 60 to 80 ft. high, hardy throughout. Europe.

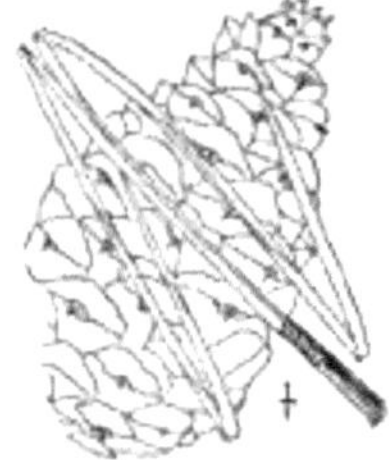

P. Larício.

12. **Pìnus Larício**, Poir. (Corsican Pine.) Leaves 4 to 6 in. long, slender, very wavy, dark green; 2 together in a sheath. Cones 2 to 3 in. long, conical, somewhat curved, often in pairs. Scales with very small prickles. Seeds rather large with broad wings. A tall, open, pyramidal, rapid-growing tree, 60 to 100 ft. high, with the branches in regular whorls, spreading and very resinous. Often cultivated. Europe.

P. Massoniàna.

13. **Pìnus Massoniàna**, Sieb. (Masson's Pine.) Leaves in twos, 4 to 6 in. long, rather stiff, concave on one side and convex on the other, twisted but not curved; sharp-pointed, of a fresh, [Pg 176] bright green color. Cones 1 to 1½ in. long, conical, incurved, solitary but numerous, with closely overlapping scales terminating in slender prickles. An upright, compact tree, 40 to 50 ft. high, from Japan; sometimes cultivated. Hardy at Boston.

P. mìtis.

14. **Pìnus mìtis**, Michx. (Common Yellow Pine.) Leaves sometimes in threes, usually in twos, from long sheaths; slender, 3 to 5 in. long, dark green, rather soft. Cones ovate to oblong-conical, hardly 2 in. long; the scales with minute weak prickles. A large tree with an erect trunk, 50 to 100 ft. high. Staten Island, south and west. The western form has more rigid leaves, and more spiny cones.

P. densiflòra.

15. **Pìnus densiflòra**, Siebold. (Japan Pine.) Leaves about 4 in. long, from short, fringed, scale-like sheaths; rigid, convex above, concave beneath and somewhat serrulate on the margin, very smooth, sharp-pointed and crowded, shining green and somewhat glaucous; falling when one to two years old; 2 in a sheath. Cones abundant; 1½ in. long, short-peduncled, conical, obtuse, terminal, somewhat pendent; scales linear-oblong, woody, with a small prickle which soon falls off. A beautiful small tree, 30 to 40 ft. high; from Japan; hardy throughout.

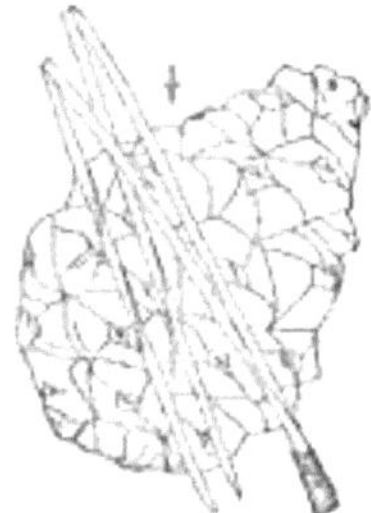

P. resinòsa.

16. **Pìnus resinòsa**, Ait. (Red Pine.) Leaves 5 to 6 in. long, in twos, from long sheaths; rigid, straight, dark green. Cones 2 in. long, ovate-conical, smooth, their scales without points, slightly thickened, usually growing in clusters. A tall tree, 60 to 80 ft. high, with rather smooth, reddish bark and hard light-colored wood; branchlets also having smooth reddish bark. Pennsylvania, north and west.
[Pg 177]

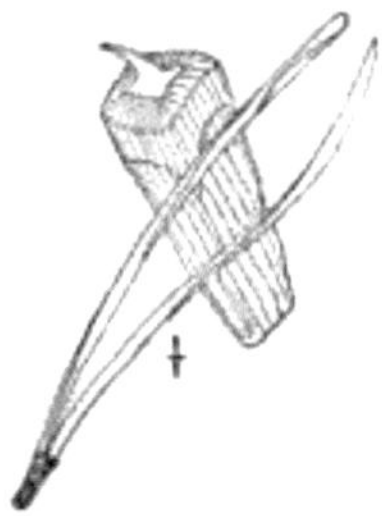

P. ínops.

17. **Pìnus ínops**, Ait. (Jersey or Scrub Pine.) Leaves short, 1½ to 3 in. long, rigid; usually 2, rarely 3, in a short sheath. Cones solitary, 2 to 3 in. long, ovate-oblong, curved, on a short stalk. Scales tipped with a straight, rigid spine. A small tree, 15 to 30 ft. high, growing wild in sections where the soil is poor and sandy; having straggling flexible branches with rough, dark bark; New Jersey, south and west. Rarely cultivated.

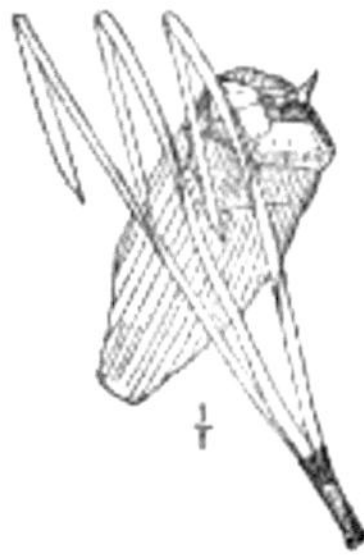

P. púngens.

18. **Pinus púngens**, Michx. f. (Table-Mountain Pine.) Leaves in twos, sometimes in threes, stout, short, 1¼ to 2½ in. long, crowded, bluish; the sheath short (very short on old foliage). Cones 3 in. or more long, hanging on for a long time; the scales armed with a stout, hooked spine, ¼ in. long. A rather small tree, 20 to 60 ft. high. New Jersey and south westward, along the mountains.

P. sylvéstris.

19. **Pìnus sylvéstris**, L. (Scotch Pine, wrongly called Scotch Fir.) Leaves in twos, 1½ to 2½ in. long, from short, lacerated sheaths, twisted, rigid, of a grayish or a glaucous-green color. Cones 2 to 3 in. long, ovate-conical, of a grayish-brown color, ripening the second year, the scales having 4-sided, recurved points. A large and very valuable tree of central Europe. Many varieties are in cultivation in this country. It forms the Red and Yellow Deal so extensively used for lumber in Europe.

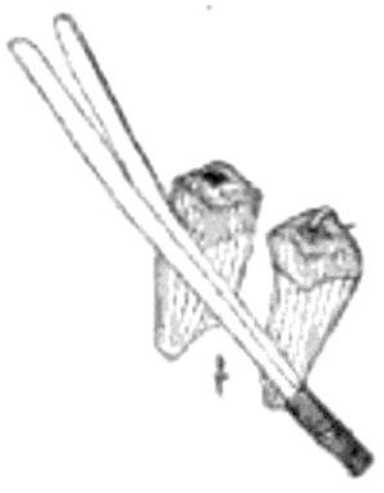

P. contórta.

20. **Pìnus contórta**, Dougl. (Twisted-branched Pine.) Leaves 2 in. long, nu [Pg 178] merous, rigid, sharply mucronate, from a short, dark, overlapping sheath; 2 to a sheath. Cones from 2 to 2½ in. long, ovate, smooth, clustered. Scales furnished with a point which is soon shed. A small cultivated tree, 30 to 40 ft. high, from the Pacific coast of the United States. As it has an irregular shape, and crooked branches, it is not often planted.

P. Banksiàna.

21. **Pìnus Banksiàna**, Lambert. (Gray or Northern Scrub Pine.) Leaves in twos, short, 1 in. long, oblique, divergent from a close sheath. Cones lateral, conical, oblong, usually curved, 1½ to 2 in. long, the scales thickened at the end and without points. A straggling shrub, sometimes a low tree, found wild in the extreme Northern States.

P. édulis.

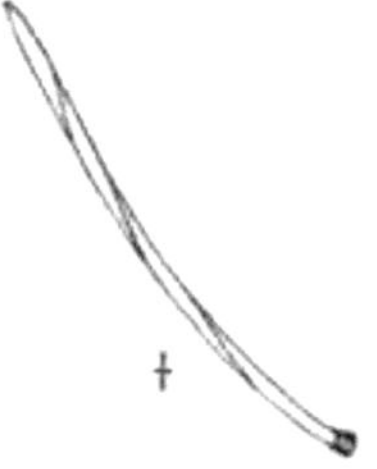

P. monophýlla.

22. **Pìnus édulis**, Engelm. (Piñon or Nut-pine.) Leaves mostly in pairs, rarely in threes, 1 to 1½ in. long, from short sheaths, light-colored, rigid, curved or straightish, spreading; cones sessile, globose or nearly so, 2 in. long; tips of scales thick, conical-truncate, no awns or prickles; seeds large, nut-like, wingless, edible. A low, round-topped tree, branching from near the base, 10 to 25 ft. high;

from the Rocky Mountains. A fine small pine; cultivated in the East. It needs some protection at Boston. The figure shows the seed. **Pìnus monophýlla**, Torr. and Frem., from the mountain regions farther west, has its leaves in ones and twos; when in ones, round and very rigid; when in pairs, flat on the inner side; leaves on the young shoots bluish, glaucous green, or silvery. This is probably only a variety of P. edulis. The seeds of both are so large and nutritious that they are extensively used for food by the Indians. [Pg 179]

Genus **94. PÌCEA.** (The Spruces.)

Leaves evergreen, scattered (pointing in every direction), needle-shaped, keeled above and below, thus making them somewhat 4-sided. Fertile catkins and cones terminal; cones maturing the first year, pendulous; scales thin, without prickles, persistent, the cone coming off the tree whole.

* Leaves very short, usually ¼ to ½ in. long, obtuse   7, 8.

* Leaves usually ½ in. or more long, acute. (**A.**)

   A. Cones over 3 in. long; cultivated. (**B.**)

      B. Leaves dark green; large tree, common   3.

      B. Leaves bright or pale green   4, 5, 6.

   A. Cones 2 in. or less long; large native trees   1, 2.

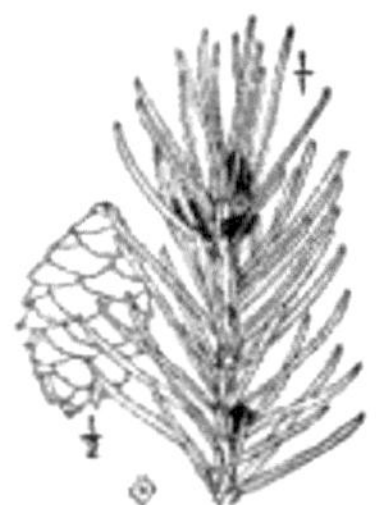

P. nìgra.

1. **Pìcea nìgra**, Link. (Black or Double Spruce.) Leaves about ½ in. long, erect, stiff, somewhat 4-sided, very dark green or whitish-gray; branchlets pubescent. Cones persistent, 1 to 1½ in. long, ovate or ovate-oblong, changing from dark purple to dull reddish-brown; scales very thin, roundish, with toothed or uneven edges. A conical-shaped tree, 40 to 80 ft. high; wild in the North and along the Alle-

ghanies; often cultivated. Bark dark brown; branches horizontal; wood light reddish.

Var. *rubra* has larger, darker leaves, and larger, brighter-colored cones.

P. álba.

2. **Pìcea álba**, Link. (White or Single Spruce.) Leaves ½ to ¾ in. long, rather slender, needle-shaped, sharp-pointed, incurved, pale- or glaucous-green; branchlets smooth. Cones deciduous, 2 in. long, oblong-cylindrical, with entire, thin-edged scales. Tree 25 to 100 ft. high, of beautiful, compact, symmetrical growth when young, and such light-colored foliage as to make it a fine [Pg 180] species for cultivation. Wild in the North, and cultivated throughout. There are varieties with bluish-green (var. *cærulea*) and with golden (var. *aurea*) foliage in cultivation.

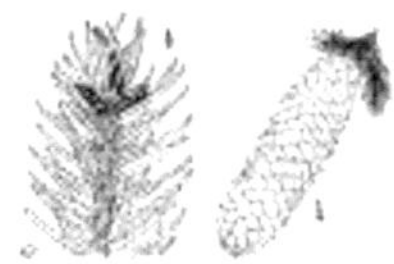

P. excélsa.

3. **Pìcea excélsa**, Link. (Norway Spruce.) Leaves ¾ to 1 in. long, rigid, curved, dark green. Cones 5 to 7 in. long, and pendent at maturity, with the scales slightly incurved. A large tree, 70 to 120 ft. high, of vigorous growth, with numerous, stout, drooping branches; abundant in cultivation. A score of named varieties are sold at the nurseries, some quite dwarf, others so very irregular in shape as to be grotesque.

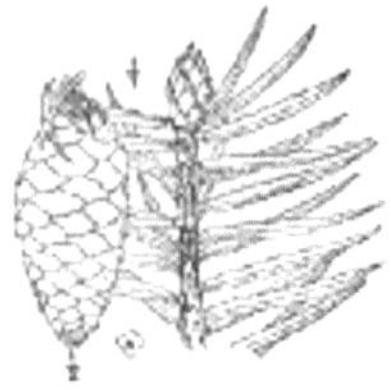

P. políta.

4. **Pìcea políta**, Carr. (Tiger's-tail Spruce.) Leaves ½ to ¾ in. long, strong, rigid, sharp-pointed, somewhat curved, glabrous, bright green, on stout branches with prominent buds. Leaves persistent for 7 years; not 2-ranked. Cones 4 to 5 in. long, spindle-shaped elliptical, rounded at the ends. Tree of slow growth, with horizontal, yellowish-barked branches. As it is a tree of recent introduction (1865) from Japan, there are no large specimens. Hardy at Boston. [Pg 181]

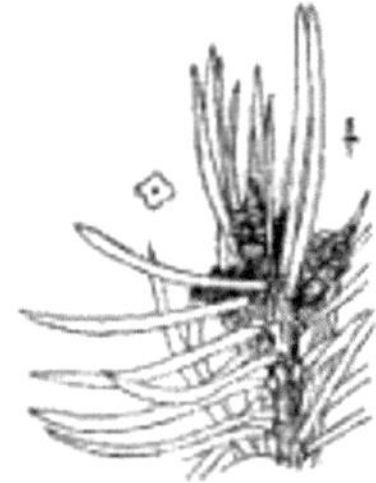

P. púngens.

5. **Pìcea púngens**, Eng. (Silver Spruce.) Leaves ½ to 1 in. long, broad, rigid, stout, sharply acute, usually curved, pale green above, silvery-glaucous beneath, on smooth and shining branchlets. Cones very abundant, 3 to 5 in. long, cylindric, with elongated, undulated, retuse scales. A strictly conical tree with spreading branches and thick, smooth, gray bark. Sometimes cultivated; from the Rocky Mountains. Hardy.

P. Morínda.

6. **Pìcea Morínda**, Link. (Himalayan Spruce.) Leaves 1 to 2 in. long, very sharply acute, pale green color, spreading, 4-sided, straight, rigid, slightly glaucous beneath; branches horizontal; branchlets remotely verticillate, numerous, drooping, with light-colored bark. Cones 6 to 7 in. long, ovate-oblong; scales light brown, oblong, entire, smooth, loosely imbricated. A tall tree, cultivated from eastern Asia and not hardy north of Washington except in sheltered positions.

P. Alcóquina.

7. **Pìcea Alcóquina**, Lindl. (Alcock's Spruce.) Leaves ¼ to ¾ in. long, crowded, somewhat 4-sided, flattish, recurved, obtusely rounded at tip, deep green above, whitish or yellowish below.

243

Cones 2 to 3 in. long, 1 in. in diameter, reddish fawn-color, with very persistent scales; scales wedge-shaped at base, rounded at tip. A large tree from Japan; fully hardy as far north as Mass.

P. orientàlis.

8. **Pìcea orientàlis**, L. (Eastern Or Oriental Spruce.) Leaves very short, ½ in. long, [Pg 182] 4-sided, rigid, stout, rather obtuse, dark shining green, entirely surrounding the branches. Cones 2½ to 3 in. long, cylindrical, with soft, thin, loose, rounded scales, uneven on the edges. A beautiful, conical, slow-growing, compact tree, reaching the height of 75 ft.; often cultivated; from the Black Sea. Hardy.

Genus **95. TSÙGA.** (Hemlocks.)

Leaves evergreen, scattered, flat, narrowed to a green petiole, appearing 2-ranked by the direction they take, whitened beneath. Fertile catkins and cones on the end of last year's branchlets. Cones pendulous, maturing the first year; scales thin, persistent.

T. Canadénsis.

1. **Tsùga Canadénsis**, Carr. (Common Hemlock.) Leaves short-petioled, linear, ½ in. long, obtuse, dark green above and white beneath; the young leaves in the spring a very light green. Cones oval, ½ to ¾ in. long, pendent, of few (20 to 40) scales. A large, very beautiful tree, 50 to 80 ft. high, abundant in rocky woods, and cultivated throughout; spray light and delicate.

T. Caroliniàna.

2. **Tsùga Caroliniàna**, Engelm. (Mountain-hemlock.) This is similar to the last; its leaves are larger, glossier, more crowded; its cones are larger, and have wider and more spreading scales; the tree is smaller, rarely growing 40 ft. high. Wild, but scarce, in the higher Alleghanies, south; beginning to be cultivated north, and probably hardy throughout.

T. Siebòldii.

3. **Tsùga Sieboldii.** (Japan Hemlock.) Leaves ½ to ¾ in. long, linear, [Pg 183] obtuse to notched at the tip, smooth, thick, dark green above, with two white lines below. Cones scarcely 1 in. long, elliptical, solitary, terminal, obtuse, quite persistent; scales pale brown. A beautiful small tree, 20 to 30 ft. high, with an erect trunk, dark-

brown bark, and numerous, pale, slender branchlets. Introduced from Japan, and probably hardy throughout.

Genus **96. ÀBIES.** (The Firs.)

Leaves evergreen, flat, scattered, generally whitened beneath, appearing somewhat 2-ranked by the directions they take. Fertile catkins and cones erect on the upper side of the spreading branches. Cones ripening the first year; their scales thin and smooth, and the bracts generally exserted; scales and bracts breaking off at maturity and falling away, leaving the axis on the tree. A great number of species and varieties have been planted in this country, but few if any besides those here given do at all well in our dry and hot climate.

* Cones 6 to 8 in. long; leaves blunt at tip. (**A.**)

    A. Leaves over an inch long   10, 11.

    A. Leaves an inch or less long   12.

* Cones 3½ to 6 in. long. (**B.**)

    B. Leaves 2 in. or more long, 2-ranked   9.

    B. Leaves 1 in. or less long. (**C.**)

        C. Leaves acute at tip   7, 8.

        C. Leaves blunt or notched at tip. (**D.**)

            D. Two-ranked   4.

            D. Not 2-ranked   3.

* Cones 1 to 3½ in. long. (**E.**)

    E. Leaves an inch or more long   5, 6.

    E. Leaves less than an inch long   1, 2.

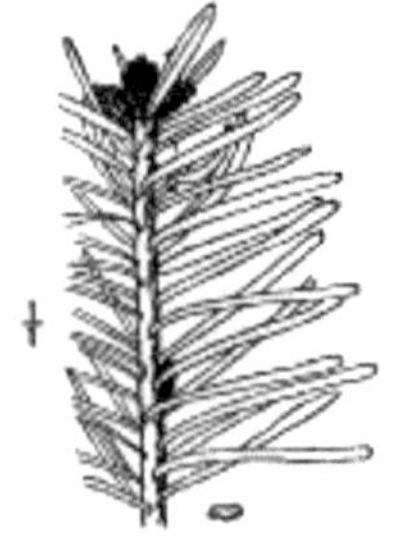

A. balsàmea.

1. **Àbies balsàmea**, Mill. (Common Balsam-fir.) Leaves narrow, linear, ½ to ¾ in. long, and much crowded, silvery beneath; those on the horizontal branches spreading into 2 ranks. Bark yielding Canada balsam from blisters. Cones erect, on spreading branches, 2 to 4 in. long and 1 in. thick, cylindric, violet-col [Pg 184] ored, with mucronate-pointed bracts extending beyond the scales and not reflexed. Wild in cold, wet grounds; 20 to 45 ft. high, with numerous horizontal branches. Has been cultivated quite extensively, although there are better Firs for ornamental purposes.

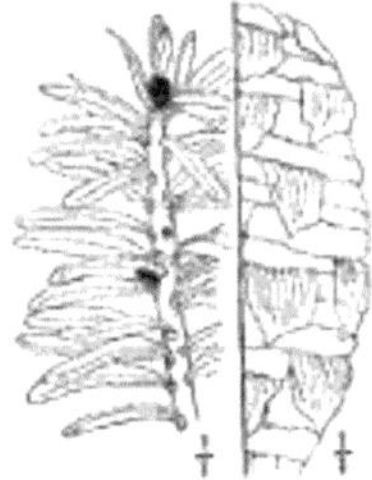

A. Fràseri.

2. **Àbies Fràseri**, Lindl. (Fraser's or Southern Balsam-fir.) Leaves ½ to ¾ in. long, somewhat 2-ranked, linear, flattened, obtuse, emarginate, whitish beneath, the lower ones curved and the upper ones erect. Cones oblong, 1 to 2 in. long, with sharp-pointed bracts half exserted and reflexed. A rare, small tree, 30 to 40 ft. high, growing wild in the mountains, from Virginia south. A hardy tree and handsome when young.

A. Nordmanniàna.

3. **Àbies Nordmanniàna**, Link. (Nordmann's Silver Fir.) Leaves very numerous, crowded, broad, linear, blunt or erose-dentate at the ends, somewhat curved, of unequal length, 1 in. or less long, deep green above and whitened beneath. Cones large, 5 in. long, ovate, erect, with very obtuse scales; bracts exserted and recurved. A beautiful large tree, 50 to 80 ft. high, occasionally cultivated; with numerous horizontal branches and smooth bark.

A. fírma.

4. **Àbies fírma**, S. and Z. (Japan Silver Fir.) Leaves ¾ to 1 in. long, very closely 2-ranked, slightly twisted, linear, somewhat notched at the end, smooth and dark above, somewhat silvery below. Cones 3 to 4½ in. long, 1 to 1½ in. [Pg 185] in diameter, straight, cylindric, with broad, downy, leathery, crenulated scales; bracts exserted, with acute, slightly recurved points. A beautiful tall tree with somewhat the habit of the common Silver Fir; recently introduced from Japan, and hardy as far north as central New York.

A. grándis.

5. **Àbies grándis**, Lindl. (Great Silver Fir.) Leaves 1 to 1½ in. long, mostly curved, deep green above and silvery below, not 2-ranked. Cones 3 in. long and about 2 in. broad, obtuse, solitary, chestnut-brown in color. A very large (200 to 300 ft. high), handsome tree

from the Pacific coast. Hardy at Washington; needs protection north.

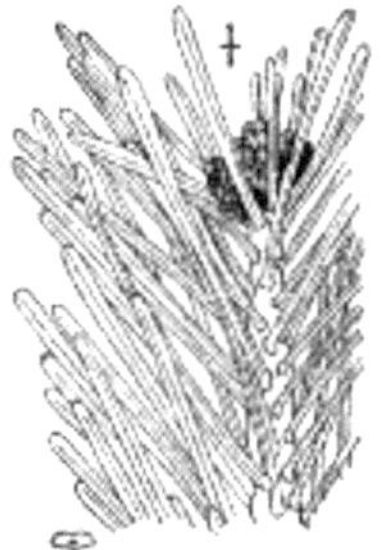

A. Píchta.

6. **Àbies Píchta**, Fisch. (Siberian Silver Fir.) Leaves 1 in. long, linear, flat, obtuse, incurved at the apex, mostly scattered, very dark green above, paler beneath. Cones 3 in. long, ovate, cylindric, obtuse, with rounded, entire scales and hidden bracts. A small to medium-sized cultivated tree, 25 to 50 ft. high, with horizontal, somewhat pendulous branches and dense compact growth. It is peculiar in its very dark foliage; very hardy.

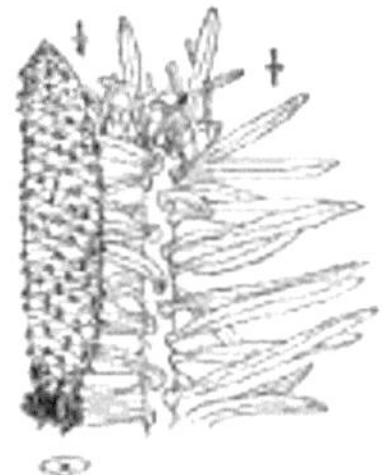

A. Cephalónica.

7. **Àbies Cephalónica**, Loud. (Cephalonian Silver Fir.) Leaves ¾ in. long, very stiff, sharp-pointed, spreading broadly from the branches in all directions, dark green above and white beneath; petioles very short, dilated lengthwise at the point of attachment of the branches. Cones very erect, 4 to 6 in. long, 1-1/3 in. in diameter; projecting scales unequally toothed and reflexed at the point. A beautiful, cultivated [Pg 186] tree, 30 to 60 ft. high, with bright brown bark and resinous buds.

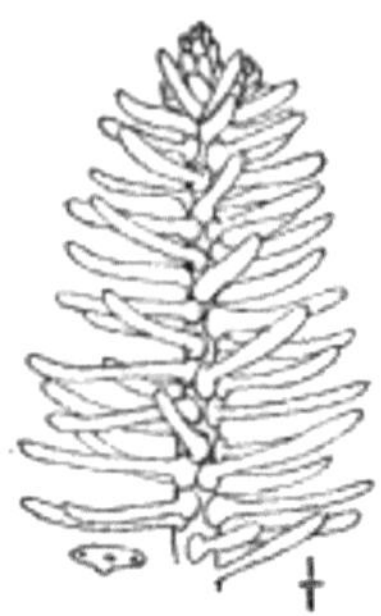

A. Pinsàpo.

8. **Àbies Pinsàpo**, Bois. (Pinsapo Fir.) Leaves less than 1 in. long (usually ½ in.), rigid, straight, scattered regularly around the branches, and pointing in all directions; disk-like bases large; branches in whorls, and branchlets very numerous. Cones 4 to 5 in. long, oval, sessile; scales rounded, broad, entire; bracts short. A very handsome tree from Spain, and reported hardy at the Arnold Arboretum.

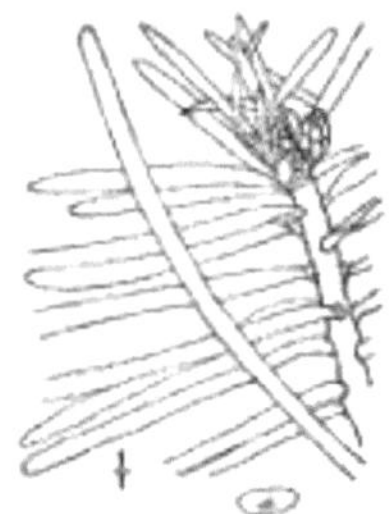

A. cóncolor.

9. **Àbies cóncolor**, Lindl. (White Fir.) Leaves 2 to 3 in. long, mostly obtuse, but on young trees often long-pointed, 2-ranked, not crowded on the stem, pale green or silvery. Cones oblong-cylindric, 3 to 5 in. long, 1½ in. in diameter; scales twice as broad as long; bracts short, not projecting. A large tree, 75 to 150 ft. high; bark rough, grayish. Native in the Rocky Mountains; hardy at the Arnold Arboretum, Massachusetts, but needs some protection at St. Louis.

A. Cilícica.

10. **Àbies Cilícica**, Carr. (Cilician Silver Fir.) Leaves flat, linear, 1 to 1¾ in. long and 1/12 in. broad, somewhat 2-ranked but rather irregularly scattered around the young shoots; shining dark green above and whitish beneath. Cones 7 to 8 in. long, nearly 2 in. in diameter, cylindric, obtuse, erect, with thin and entire scales, and short and hidden bracts. A very conical tree, 50 ft. high, with branches in whorls, and numerous, small, slender branchlets. Bark light gray; recently cultivated from Asia. [Pg 187]

A. nóbilis.

11. **Àbies nóbilis**, Lindl. (Noble Silver Fir.) Leaves 1 to 2 in. long, linear, much curved, the base extending a short distance upward along the branch, then spreading squarely from it, crowded, compressed, deep green above, glaucous below; base of the leaf much less disk-like than in most of the Firs; branches horizontal, spreading, numerous. Cones 6 to 7 in. long and nearly 2 in. in diameter, cylindric, sessile, with large, entire, incurved scales; bracts large, exserted, reflexed, spatulate, with terminal, awl-shaped points. A very large, beautiful tree, from the Pacific coast, where it grows 200

ft. high. Hardy in Pennsylvania, but needs some protection in Massachusetts.

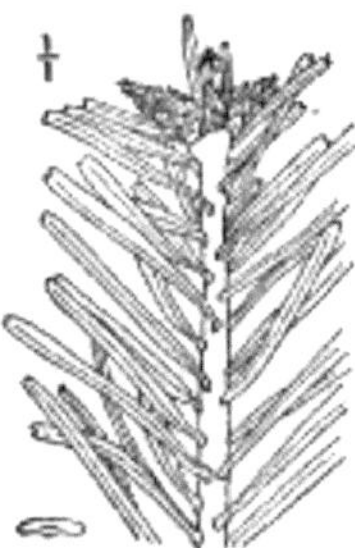

A. pectinàta.

12. **Àbies pectinàta**, DC. (European or Common Silver Fir.) Leaves ½ to 1 in. long, linear, obtuse, occasionally with an incurved point, polished green above, two white lines below, rigid, straight; branches horizontal and in whorls. Cones 6 to 8 in. long, cylindric, brown when ripe; scales broad, thin, rounded; bracts long, exserted, with an acute reflexed tip. Introduced from Europe. Good specimens can be found as far north as Massachusetts, though our climate is not fitted to give them either long life or perfect form.

Genus **97. LÀRIX.** (The Larches.)

Leaves deciduous, all foliaceous, the primary ones scattered, but most of them in bundles of numerous leaves from lateral globular buds. Cones usually small (in one cultivated species 3 in. long), ovoid, erect, with smooth scales.

* Cones less than 1 in. long, of not more than 25 scales	1.
* Cones 1 to 2 in. long, of from 40 to 60 scales	2, 3.
* Cones 2 to 3 in. long, with thick, woody, somewhat divergent deciduous scales. (Pseudolarix)	4.

[Pg 188]

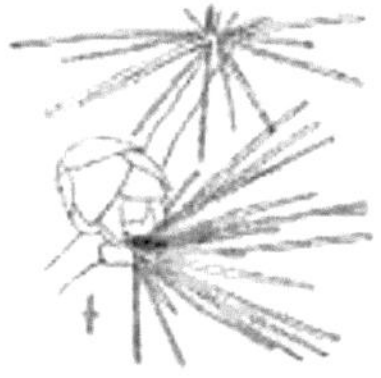

L. Americàna.

1. **Làrix Americàna**, Michx. (American Larch. Tamarack or Hackmatack.) Leaves less than 1 in. long, thread-like, linear, slender, light bluish-green. Cones ½ to ¾ in. long, ovoid, of a reddish color. A tree of large size, 50 to 100 ft. high, growing wild in all the northern portion of our region, and frequent in cultivation, although not quite so fine a tree as Larix Europæa.

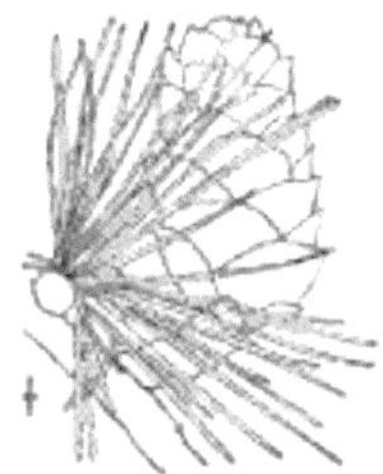

L. Europæ̀a.

2. **Lárix Europæ̀a**, DC. (European Larch.) Leaves 1 in. long, linear, obtuse, flat, soft, numerous, and bright green in color. Cones sometimes more than 1 in. long, with oval, erect, very persistent scales. A beautiful tree with horizontal branches and drooping branchlets; abundant in cultivation.

Var. *pendula* has long, pendent branches, and forms a very fine weeping tree.

L. Leptolépsis.

3. **Làrix Leptolépsis**, Gordon. (Japan Larch.) Leaves 1 to 1½ in. long, slender, pale green. Cones 1-1/3 in. long, and half as wide, of about 60 scales, reflexed at the margin, pale brown in color; bracts lanceolate, acute, entire, thin, one half the length of the scales; seeds obovate, compressed, with long, obtuse, thin wings. A small tree from northern Japan, where it grows 40 ft. high. It is a handsome, erect-growing tree, with slender, smooth, ash-colored branches, and rather rigid, spreading branchlets. [Pg 189]

L. Kæmpferi

4. **Làrix Kæmpferi**, Lamb. (Golden Larch.) Leaves from 1 to 2½ in. long, flat, linear, sword-shaped, somewhat soft, pale pea-green in the spring, golden-yellow in the autumn. Cones 2 to 3 in. long, with flattish, divergent scales which are very deciduous. A beautiful large tree, over 100 ft. high, from China, which proves hardy as far north as central New York. It is often placed in a new genus (Pseudolarix) because of the deciduous scales to the cones.

Genus **98. CÈDRUS.** (The Lebanon Cedars.)

Leaves linear, simple, evergreen, in large, alternate clusters. Cones large, erect, solitary, with closely appressed scales; seeds adhering to the base of their lacerated, membranous wings. Large, spreading-branched trees from southern Asia and northern Africa. Occasionally successfully grown from New York City southward.

* Leaves 1 in. or less long	1, 2.
* Leaves over 1 in. long, light glaucous-green	3.

C. Libàni.

1. **Cèdrus Libàni**, Barr. (Cedar of Lebanon.) Leaves ¾ to 1 in. long, acuminate, needle-form, rigid, few in a fascicle, deep green in color. Cones 3 to 5 in. long, oval, obtuse, very persistent, grayish-brown in color; scales thin, truncate, slightly denticulate; seeds quite large and irregular in form. A cultivated tree with wide-spreading, whorled, horizontal branches covered with rough bark. Somewhat tender when young in the Middle States, but forming a grand tree in proper positions. [Pg 190]

C. Atlántica.

2. **Cèdrus Atlántica**, Manetti. (Mt. Atlas, Silver, or African Cedar.) Leaves ½ to ¾ in. long, mostly cylindric, straight, rigid, mucronate, crowded, and of a beautiful glaucous-green color. Cones 2½ to 3 in.

long, ovate, glossy. This beautiful tree has been considered a silvery variety of Cedrus Libani. They are about alike in hardiness and in general form. Cedrus Atlantica has more slender branches, denser and more silvery foliage. From Africa.

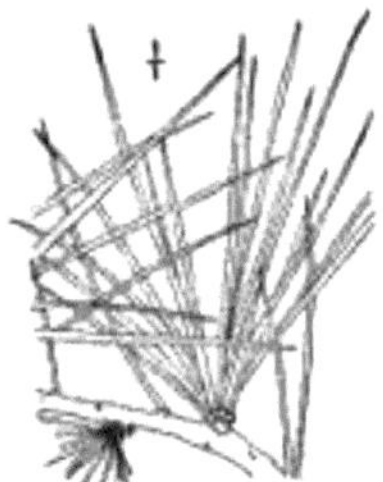

C. Deodàra.

3. **Cèdrus Deodàra**, Lindl. (Deodar or Indian Cedar.) Leaves 1 to 2 in. in length, 3- or usually 4-sided, rigid, acute, very numerous (about 20 in a fascicle), bright green, covered with a glaucous bloom. Cones 4 to 5 in. long, ovate, obtuse, very resinous, rich purple when young, and brown when old; the scales separating from the axis at maturity. Seeds wedge-shaped, with large, bright brown wings. A beautiful pyramidal tree, with graceful drooping branches and light silvery foliage. Not hardy north of Philadelphia; from India.

Genus **98a. ARAUCÀRIA.**

A. imbricàta.

Araucària imbricàta, Pavon. (Chile Pine.) Leaves 1 to 2 in. long, ovate-lanceolate, sessile, rigid, acute, very persistent, closely overlapping, completely covering the thick stems, in whorls of 6 to 8, deep glossy green; branches horizontal, in whorls of 6 to 8, with ascending tips, covered with resinous, corky bark. Flowers dioecious; cones (on only a portion of [Pg 191] the trees) large, roundish, about 7 in. in diameter, erect, solitary; seeds wedge-shaped, 1 to 2 in. long. A large, peculiar, beautiful, conical tree, with much the appearance of a cactus; not fitted to our climate, although a few specimens may be found growing quite well near the coast south of Philadelphia. From the mountains of Chile.

Genus **99. CUNNINGHÀMIA.**

A genus of but one species. The cone-scales are very small, but the bracts are large, thick, and serrate.

Cunninghàmia Sinénsis, R. Br. (Cunninghamia.) Leaves 1½ to 2½ in. long, flat, rigid, numerous, alternate, somewhat serrulate; the leaf gradually increases in width from the acute tip to the base, which is decurrent on the stem and about 1/8 in. wide. Cones 1 to 1½ in. long, nearly globular, erect, very persistent, mostly clustered, sessile; the scale is a mere transverse ridge, but the bract is large and prominent, like a triangular-hastate, dilated leaf. A very handsome tree, from China, which does not succeed very well in this region except in protected situations.

Genus **100. SCIADÓPITYS.**

Cones elliptical or cylindrical, large, obtuse. Leaves evergreen, somewhat flattened, arranged in distant whorls around the stems, and spreading in all directions.

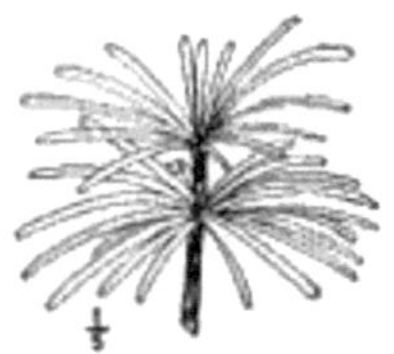

S. verticillàta.

Sciadópitys verticillàta, S. and Z. (Umbrella-pine.) Leaves 2 to 4 in. long, 1/6 in. wide, linear, obtuse, smooth, persistent, sessile, entire, in whorls of 30 to 40 at the nodes and extremity of the branches. Cones 3 by 1½ in. Scales wedge-shaped, corrugated, overlapping, coriaceous, persistent; bracts adherent, broad, and smooth. A beautiful, tall, conical, slow-growing tree, with the branches whorled. Recently introduced; hardy in the New England States. [Pg 192]

Genus **101. TAXÒDIUM.**

Leaves deciduous, spreading, in 2 ranks. Flowers monœcious on the same branch, the staminate ones in spikes, and the pistillate ones in pairs below. Cones globular; the scales peltate, angular, thick, firmly closed till ripe, with 2 angular seeds under each.

T. dístichum.

Taxòdium dístichum, Richard. (Southern or Bald Cypress.) Leaves deciduous, flat, linear, ½ to ¾ in. long, in 2 rows on the slender branchlets, forming feather-like spray of a light green color. This whole spray usually falls off in the autumn as though a single leaf. Cones round, closed, hard, 1 in. in diameter. A fine, tall (100 to 125 ft. high), slender, spire-shaped tree with a large, spreading, rigid trunk, 6 to 9 ft. thick, and peculiar conical excrescences (called knees) growing up from the roots. Wild from Maryland south, and cultivated and hardy in the Middle and many of the Northern States.

Var. pendulum.

Var. *pendulum*, with horizontal branches and drooping branchlets, has the leaves but slightly spreading from the stems, especially when young. Very beautiful; hardy as far north as Massachusetts.

Genus **102. SEQUÓIA.**

Flowers monœcious, terminal, solitary, catkins nearly globular. Seeds winged, 3 to 5 under each scale.

S. gigántea.

1. **Sequóia gigántea**, Torr. (Big or Great Tree of California.) Leaves on the young shoots spreading, needle-shaped, sharp-pointed, scattered spirally around the branchlets; finally [Pg 193] scale-shaped, overlapping, mostly appressed, with generally an acute apex, light green in color. Cones oval, 2 to 3 in. long, of about 25 scales. The largest tree known, 300 ft. high, with a trunk nearly 30 ft. through, found in California and occasionally planted east, though with no great success, as it is almost certain to die after a few years.

S. sempérvirens.

2. **Sequóia sempérvirens**, Endl. (Redwood.) Leaves from ½ to 1 in. long, linear, smooth, 2-ranked, flat, acute, dark shining green, glaucous beneath; branches numerous, horizontal, spreading. Cones 1 in. long, roundish, solitary, terminal; scales numerous, thick, rough, furnished with an obtuse point. A magnificent tree from California, where it grows 200 to 300 ft. high. In the East it can be kept alive but a few years even at Washington.

Genus **103. THÚYA.** (Arbor-vitæ.)

Small, evergreen trees with flat, 2-ranked, fan-like spray and closely overlapping, small, appressed leaves of two shapes on different branchlets, one awl-shaped and acute, the other scale-like, usually blunt and close to the branch. Fertile catkins of few, overlapping scales fixed by the base; at maturity, dry and spreading. There are scores of named varieties of Arbor-vitae sold by the nurserymen under 3 different generic names, Thuya, Biota, and Thuyopsis. There are but slight differences in these groups, and they will in this work be placed together under Thuya. Some that in popular language might well be called Arbor-vitæ (the Retinosporas) will, because of the character of the fruit, be included in the next genus.

* Scales of the cones pointless, thin, straight. (Thuya)   1, 2.

* Scales reflexed and wedge-shaped. (Thuyopsis)   3.

* Scales thick, with horn-like tips. (Biota)   4.

[Pg 194]

260

T. occidentàlis.

1. **Thùya occidentàlis**, L. (American Arbor-vitæ. White Cedar.) Leaves in 4 rows on the 2-edged branchlets, having a strong aromatic odor when bruised. Cones oblong, 1/3 in. long, with few (6 to 10) pointless scales. A small tree, 20 to 50 ft. high, or in cultivation 1 to 50 ft. high, with pale, shreddy bark, and light, soft, but very durable wood. Wild north, and extensively cultivated throughout under more than a score of named varieties. Their names—*alba, aurea, glauca, conica, globosa, pyramidalis, pendula,* etc.—will give some idea of the variations in color, form, etc.

T. gigantèa.

2. **Thùya gigantèa**, Nutt. (Giant Arbor-vitæ.) Leaves scale-shaped, somewhat 4-sided, closely overlapping, sharp-pointed, slightly tuberculate on the back; cones more or less clustered and nearly ½ in. long. A very large and graceful tree, 200 ft. high, with white, soft wood; from the Pacific coast; introduced but not very successfully grown in the Atlantic States.

T. dolabràta.

3. **Thùya dolabràta**, L. (Hatchet-leaved Arbor-vitæ.) Leaves large, sometimes ¼ in. long, very blunt, in 4 rows on the flattened spray. Cones quite small, ovate, sessile, with jagged edges; scales reflexed and wedge-form. A small conical tree with horizontal branches and drooping branchlets; which, because of its large leaves (for an Arbor-vitæ) and flexible branchlets, is quite unique and interesting. In shaded and moist places it has done quite well as far north as New York.

T. orientàlis.

4. **Thùya orientàlis**, L. (Eastern or Chinese Arbor-vitæ.) Leaves small, in 4 opposite rows, appressed, acute, on the numerous 2-edged branchlets. Cones large, roundish, with thick leathery [Pg 195] scales having recurving, horn-like tips. Of this species there are as many varieties sold as of number one, and nearly the same varietal names are used; but it is not so good a species for general cultivation in this country.

Var. *flagelliformis*, Jacq. (Weeping Arbor-vitæ), has very slender, elongated, weeping branches, curving gracefully to the ground. It is a beautiful variety, often cultivated (a single stem is shown in the figure).

Genus **104. CHAMÆCÝPARIS.** (The Cypresses.)

Strong-scented, evergreen trees with very small, scale-like or somewhat awl-shaped, closely appressed (except in some cultivated varieties), overlapping leaves and 2-ranked branchlets, almost as in Thuya. Cones globular, with peltate, valvate scales, firmly closed till ripe; the scales thick and pointed at the center.

* Native trees; leaves light glaucous-green.   1.

* Cultivated trees from Western America; leaves dark green. (**A.**)

  A. No tubercle on the backs of the leaves.   2.

  A. Usually a tubercle on the back   3.

* Cultivated small trees and shrubs from Japan (called Retinospora)   4.

C. sphæroídea.

1. **Chamæcýparis sphæroídea**, Spach. (White Cedar.) Leaves very small, triangular, awl-shaped, regularly and closely appressed in 4 rows, of a light glaucous-green color, often with a small gland on the back. Cones very small, 1/3 in. in diameter, of about 6 scales, clustered. Tree 30 to 90 ft. high, wild in low grounds throughout; abundant in Middle States. With reddish-white wood and slender, spreading and drooping sprays; bark fibrous, shreddy; sometimes cultivated.

C. Nutkæ̀nsis.

2. **Chamæcýparis Nutkǽnsis**, Lambert. (Nootka Sound Cypress.) Leaves only 1/8 in. long, sharp-pointed, and closely ap [Pg 196] pressed, of a very dark, rich green color; very slightly glaucous, without tubercles on the back. Cones small, globular, solitary, with a fine, whitish bloom; scales 4, rough and terminating in a sharp straight point. Tree 100 ft. high in Alaska, and would make a fine cultivated tree for this region if it could stand our hot, dry summers.

C. Lawsoniàna.

3. **Chamæcýparis Lawsoniàna**, Park. (Lawson's Cypress.) Leaves small, deep green, with a whitish margin when young, forming with the twigs feathery-like, flat spray of a bluish-green color; leaves usually with a gland on the back. Cones scarcely ¼ in. in diameter, of 8 to 10 scales. A magnificent tree in California, and where it is hardy (in rather moist soil, New York and south) it forms one of our best cultivated evergreens. The leading shoot when young is pendulous.

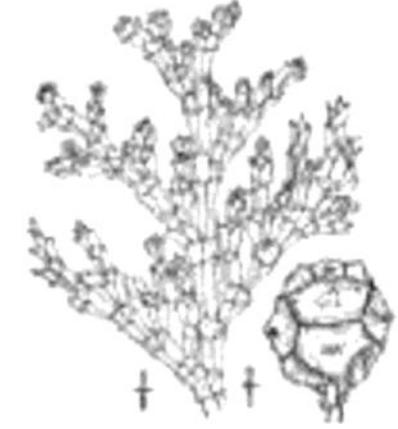

R. obtùsa.

4. **Chamæcýparis (Retinóspora) obtùsa**, Endl. (Japanese Arbor-vitæ.) Leaves scale-formed, obtuse, closely appressed and very persistent. Cones of 8 or 10 hard, light brown, wedge-shaped scales. Beautiful small trees or generally shrubs (in this country), of a score of named varieties of many colors and forms of plant and foliage.

There are probably a number of species of Japanese and Chinese Chamæcyparis (Retinospora), but till their size, hardiness, and origin have been more fully determined, it would be impossible to make an entirely satisfactory list for such a work as this. Figures are given of the common, so-called, species cultivated in this country; under each of these, several varieties are sold by the nurserymen. The three twigs of Retinospora squarrosa were all taken from a single branch; this shows how impossible it is to deter [Pg 197] mine the varieties or species; the twig at the left represents the true *squarrosa*; the others, the partial return to the original. Most of the forms shown in the figures have purple, golden, silvery, and other colored varieties.

Retinospora filifera.

Retinospora pisifera.

Retinospora squarrosa.

Retinospora Lycopoides.

[Pg 198]

Retinospora plumosa.

Genus **105. CRYPTOMÈRIA.**

A genus of evergreens containing only the following species:

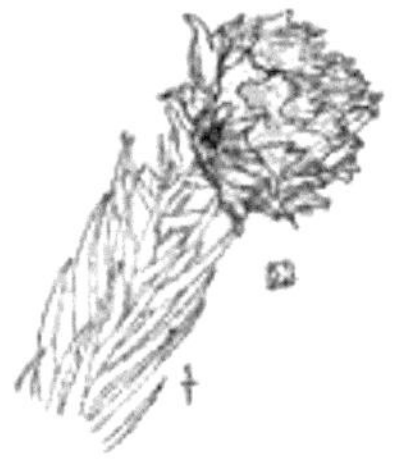

C. Japónica.

Cryptomèria Japónica, Don. (Japan Cedar.) Leaves about ½ in. long, not flattened, but about equally 4-sided, curved and tapering quite gradually from the tip to the large, sessile base; branches spreading, mostly horizontal, with numerous branchlets. Cones ½ to ¾ in. in diameter, globular, terminal, sessile, very persistent, with numerous, loose, not overlapping scales. A beautiful tree from Japan, 50 to 100 ft. high. Not very successfully grown in our climate. North of Washington, D. C., it needs a sheltered position, and should have a deep, but not very rich soil.

Genus **106. JUNÍPERUS.**

Leaves evergreen, awl-shaped or scale-like, rigid, often of two shapes on the same plant. Spray not 2-ranked. Flowers usually diœcious. Fertile catkins rounded, of 3 to 6 fleshy, coalescent scales, forming in fruit a bluish-black berry with a whitish bloom, but found on only a portion of the plants.

* Leaves rather long, ½ in., in whorls of threes 1.

* Leaves smaller; on the old branches mostly opposite 2.

J. commùnis.

1. **Juníperus commùnis**, L. (Common Juniper.) Leaves rather long, ½ in., linear, awl-shaped, in whorls of threes, prickly-pointed, upper surface glaucous-white, under surface bright green. Fruit globular, ¼ in. or more in diameter, dark purple when ripe, covered with light-colored bloom. A shrub or small tree with spreading or pendulous branches; common in dry, sterile soils. There [Pg 199] are a great many varieties of this species in cultivation, but few of them grow tall enough to be considered trees.

Var. *Hibernica* (Irish Juniper) grows erect like a column. Var. *Alpina* is a low creeping plant. Var. *hemispherica* is almost like a half-sphere lying on the ground.

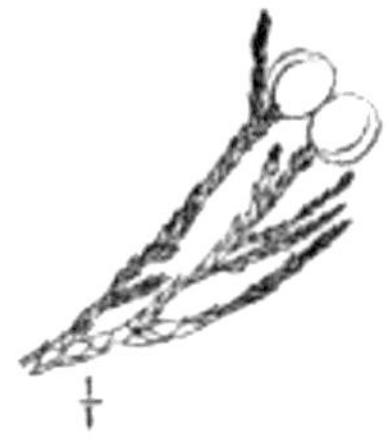

J. Virginiàna.

2. **Juníperus Virginiàna**, L. (Red Cedar.) Leaves very small and numerous, scale-like on the older branches, but awl-shaped and somewhat spreading on the young shoots; dark green. Fruit small, 1/5 in., abundant on the pistillate plants, dark purple and covered with fine, glaucous bloom. Trees from 20 to 80 ft. high (sometimes only shrubs), with mostly horizontal branches, thin, scaling bark, dense habit of growth, and dark foliage. Wood light, fine-grained, durable; the heart-wood of a handsome dark red color. Wild throughout; several varieties are found in cultivation. Many other species from China, Japan, California, etc., are occasionally cultivat-

ed, but few are large enough to be called trees, and those that are large enough are not of sufficient importance to need specific notice.

Genus **107. TÁXUS.**

Leaves evergreen, flat, linear, mucronate, rigid, scattered, appearing more or less 2-ranked. Fertile flowers and the fruit solitary; the fruit, a nut-like seed in a cup-shaped, fleshy portion formed from a disk; red.

T. baccàta.

Táxus baccàta, L. (Common European Yew.) Leaves evergreen, 2-ranked, crowded, linear, flat, curved, acute. Fruit a nut-like seed within a cup 1/3 in. in diameter; red when ripe in the autumn. As this species is somewhat diœcious, a portion of the plants will be without fruit. A widely spreading shrub rather than a tree, extensively cultivated under nearly a score of named varieties. We have a closely related wild species, **Táxus Canadénsis** (The Ground-hemlock), which is merely a low straggling bush. [Pg 200]

Genus **1O7a. TORRÈYA.**

T. taxifòlia.

The Torreyas are much like the Yews, but their leaves have two longitudinal lines, and a remarkably disagreeable odor when burned or bruised. **Torrèya taxifòlia**, Arn., from Florida, and **Torrèya Califòrnica**, Torr., from California, have been often planted. They form small trees, but probably cannot be grown successfully in the region. The figure shows a twig of T. taxifolia.

Genus 1O7b. CEPHALOTÁXUS.

C. Fortùnii.

Cephalotáxus Fortùnii, Hook., does not form a tree in this section, but a wide-spreading bush growing sometimes to the height of 10 ft., and spreading over a spot 15 ft. wide. Leaves flat, with the midrib forming a distinct ridge on both sides, linear, sometimes over 2 in. long, glossy green on the upper side, slightly whitened beneath. Fruit very large, 1 in. or more long, elliptical, with a single, thin-shelled nut-like seed covered with purplish, pulpy, thin flesh. Branches spreading, drooping, long, slender; buds small, covered with many sharp-pointed, overlapping scales; twigs green, somewhat grooved. From Japan; about hardy in New Jersey.

Genus 108. PODOCÁRPUS.

Leaves one-nerved, opposite, alternate, or scattered, linear or oblong. Flowers axillary and mostly diœcious; fruit drupe-like, with a bony-coated stone. [Pg 201]

P. Japónica.

Podocárpus Japónica, Sieb. (Japan Podocarpus.) Leaves alternate, crowded, flat, linear-lanceolate, elongated, quite sharp-pointed, narrowed to a short though distinct petiole, and continued down the stem by two ridges; leaves not 2-ranked, large, 4 to 8 in. long and ½ in. wide when growing in perfection; in specimens grown in this region, 2 to 5 in. long and ¼ in. wide; midrib forms a ridge on both sides; upper side dark glossy green; lower side with two broad whitish lines. A beautiful, erect-growing, small tree; from Japan; about hardy in central New Jersey; needs some protection in Massachusetts.

Genus **109. SALISBÙRIA.**

S. adiantifòlia.

Leaves broad, simple, alternate, stipulate, deciduous, deeply cut or lobed at the apex, alike on both surfaces, with long petioles. Flowers diœcious; staminate ones in catkins, pistillate ones either solitary or in clusters of a few each. Fruit a nut with a drupaceous covering.

Salisbùria adiantifòlia, Sm. (Ginkgo Tree.) Leaves parallel-veined, fan-shaped, with irregular lobes at the end, thick, leathery, with no midrib. Fruit globular or ovate, 1 in. long, on long, slender stems. A very peculiar and beautiful large tree, 50 to 100 ft. high;

from Japan. Hardy throughout, and should be more extensively cultivated than it is. [Pg 202]

GLOSSARY OF BOTANICAL TERMS [Pg 203]
AND
INDEX TO PART I.

The numbers refer to the pages where the illustrations appear or where fuller definitions of the words are given.

- *Abortive.* Defective or barren; not producing seeds.
- *Abrupt base of leaf*, 21.
- *Abruptly pinnate.* Pinnate, without an odd leaflet at the end; even-pinnate, 20.
- *Acerose.* Slender; needle-shaped, 20.
- *Acorn*, 27.
- *Acuminate.* Taper-pointed, 22.
- *Acute.* Terminating in a well-defined angle, usually less than a right angle, 22.
- *Adventitious buds*, 31.
- *Alternate.* Not opposite each other; as the leaves of a stem when arranged one after the other along the branch, 18.
- *Angulated.* Edge with such sudden bends as to form angles.
- *Annual layer of wood*, 13.
- *Anther.* The essential part of a stamen of a flower; the part which contains the pollen, 24.
- *Apetalous.* Said of a flower which has no corolla, 25.
- *Apex.* The point or summit, as the point of a leaf.
- *Apple-pome.* A fruit like the apple, with seeds in horny cells, 27.
- *Appressed.* Pressed close to the stem or other part, 19.
- *Ariled.* Seed with a somewhat membranous appendage, sometimes surrounding it, and attached to one end.
- *Aromatic.* With an agreeable odor.
- *Arrangement of flowers*, 26; of leaves, 18.
- *Astringent.* That which contracts or draws together muscular fiber; the opposite of laxative.
- *Auriculate.* Furnished with ear-shaped appendages, 21.
- *Awl-shaped.* Like a shoemaker's curved awl; subulate, 21.

- *Awned.* Furnished with a bristle-shaped appendage, 22.
- *Axil.* The angle between the leafstalk and the twig, 14.
- *Axillary.* Situated in the axil; as a bud, branch, or flower-cluster when in the axil of a leaf, 14, 26, 30. [Pg 204]

- *Bark*, 12.
- *Bases of leaves*, 21.
- *Berry.* Used in this work to include any soft, juicy fruit with several (at least more than one), readily separated seeds buried in the mass, 27.
- *Bipinnate.* Twice-pinnate, 20.
- *Bladdery.* Swollen out and filled with air.
- *Blade.* The thin, spreading portion, as of a leaf, 19.
- *Bract.* A more or less modified leaf belonging to a flower or fruit; usually a small leaf in the axil of which the separate flower of a cluster grows, 28.
- *Branch.* A shoot or stem of a plant, 11.
- *Branching*, general plan of, 29.
- *Branchlet.* A small branch.
- *Bristle-pointed.* Ending in a stiff, roundish hair, 22.
- *Bud.* Undeveloped branch or flower, 30; forms of, 32; bud-scales, 31.
- *Bur.* Rough-prickly covering of the seeds or fruit, 27.
- *Bush.* A shrub, 11.

- *Calyx.* The outer leafy part of a flower, 24.
- *Canescent.* With a silvery appearance, 23.
- *Capsule.* A dry, pod-like fruit which has either more than one cell, or, if of one cell, not such a pod as that of the pea with the seeds fastened on one side on a single line, 28.
- *Carpel.* That part of a fruit which is formed of a simple pistil, or one member of a compound pistil; often shown by a single seed-bearing line or part. A fruit has as many carpels as it has seed-bearing lines on its outer walls, or as it had stigmas when it was a pistil, or as it had leaves at its origin.
- *Catkin.* A scaly, usually slender and pendent cluster of flowers, 26, 28.

- *Ciliate.* Fringed with hairs along its edge.
- *Cleft.* Cut to about the middle, 22.
- *Cluster.* Any grouping of flowers or fruit on a plant, so that more than one is found in the axil of a leaf, or at the end of a stem, 26.
- *Complete.* Having all the parts belonging to an organ; a *complete leaf* has blade, leafstalk, and stipules, 19; a *complete flower* has calyx, corolla, stamen, and pistil, 24.
- *Compound.* Composed of more than one similar part united into a whole; a *compound leaf* has more than one blade, 19.
- *Conduplicate.* Folded on itself lengthwise, 33.
- *Cone.* A hard, scaly fruit, as that of a pine-tree, 28.
- *Conical.* With a circular base and sloping sides gradually tapering to a point; more slender than pyramidal. [Pg 205]
- *Convolute.* In a leaf, the complete rolling from edge to edge, 34.
- *Cordate.* Heart-shaped, the stem and point at opposite ends, 21.
- *Coriaceous.* Leathery in texture or substance.
- *Corolla.* The inner, usually the bright-colored, row of floral leaves, often grown together, 24.
- *Corymb.* A flat-topped or rounded flower-cluster; in a strict use it is applied only to such clusters when the central flower does not bloom first. See *cyme*, 26.
- *Crenate.* Edge notched with rounded teeth, 22.
- *Crenulate.* Finely crenated, 22.
- *Crisped.* Having an undulated or curled edge.
- *Cross-section of wood*, 35.
- *Cuneate.* Wedge-shaped, 21.
- *Cylindric.* With an elongated, rounded body of uniform diameter.
- *Cyme.* A flat-topped flower-cluster, the central flower blooming first, 26.

- *Deciduous.* Falling off; said of leaves when they fall in autumn, and of floral leaves when they fall before the fruit forms, 23.

- *Decurrent leaf.* A leaf which extends down the stem below the point of fastening.
- *Definite annual growth*, 29.
- *Dehiscence.* The regular splitting open of fruits, anthers, etc.
- *Dehiscent.* Opening in a regular way, 27, 28.
- *Deliquescent*, 16, 29.
- *Deltoid.* Triangular, 21.
- *Dentate.* Edge notched, with the teeth angular and pointing outward, 22.
- *Denticulate.* Minutely dentate.
- *Dichotomous.* Forking regularly by twos, as the branches of the Lilac.
- *Dilated.* Spreading out; expanding in all directions.
- *Diœcious.* With stamens and pistils on different plants, 25.
- *Distichous.* Two-ranked; spreading on opposite sides in one plane; as *leaves*, 18; or *branches*, 19.
- *Divergent.* Spreading apart.
- *Divided.* Separated almost to the base or midrib, 23.
- *Drupe.* A fleshy fruit with a single bony stone. In this book applied to all fruits which, usually juicy, have a single seed, even if not bony, or a bony stone, even if the stone has several seeds, 27.
- *Dry drupe.* Used when the material surrounding the stone is but slightly fleshy, 27.
- *Duration of leaves*, 23. [Pg 206]

- *Elliptical.* Having the form of an elongated oval, 20.
- *Emarginate.* With a notched tip, 22.
- *Endogenous.* Inside-growing; growing throughout the substance of the stem, 12.
- *Entire.* With an even edge; not notched, 22.
- *Enveloping organs.* In a flower, the calyx and corolla which cover the stamens and pistil, 25.
- *Essential organs.* In a flower, the organs needed to produce seeds; the stamens and pistil, 25.
- *Evergreen.* Retaining the leaves (in a more or less green condition) through the winter and till new ones appear, 23.

- *Excurrent.* With the trunk continued to the top of the tree, 16, 29.
- *Exogenous.* Outside-growing; growing by annual layers near the surface, 11.
- *Exserted.* Projecting beyond an envelope, as the stamens from a corolla, or the bracts beyond the scales of a cone, 28.
- *Exstipulate.* Without stipules, 19.
- *Extra-axillary buds*, 30.

- *Fasciculated.* In clusters or fascicles, 18.
- *Feather-veined.* With the veins of a leaf all springing from the sides of the midrib, 20.
- *Fibrous.* Composed of fine threads or fibers.
- *Filament.* The stalk of a stamen, 24; any thread-like body.
- *Flowering.* Having flowers.
- *Flowers*, 24; clusters of, 26; kinds of, 25.
- *Folding of leaves in the bud*, 33.
- *Foliaceous.* Like a leaf in texture or appearance.
- *Footstalk.* The stem of a leaf (petiole), or the stem of a flower (peduncle).
- *Forms of leaves*, 20.
- *Fruit*, 24, 26.

- *Gamopetalous.* Same as monopetalous, 25.
- *Glabrous.* Having a smooth surface; free from hairs, bristles, or any pubescence, 23.
- *Glands.* Small cellular organs which secrete oily, aromatic, or other products. They are sometimes sunk in the leaves, etc., as on the Prickly-ash; sometimes on the surface as small projections; sometimes on the ends of hairs. The word is also used to indicate small swellings, whether there is a secretion or not.
- *Glandular.* Having glands. *Glandular-hairy.* With glandular-tipped hairs, 23.
- *Glaucous.* Covered with a fine white powder that rubs off, 23. [Pg 207]
- *Globose.* Spherical in form. *Globular.* Nearly globose.

- *Glutinous.* Covered with a sticky gum.

- *Hairy.* Having rather long hairs, 23.
- *Halberd-shaped*, 21.
- *Head.* A compact, rounded cluster of flowers or fruit, 26.
- *Heart-shaped.* Ovate, with a notched base; cordate, 21.
- *Heart-wood*, 13, 35.
- *Herbaceous.* Without woody substance in the stem; like an herb; soft and leaf-like.
- *Hybrid.* An intermediate form of plant between two nearly related species; formed by the action of the pollen of one upon the pistil of the other.

- *Imbricated.* Overlapping one another like the shingles on a roof, 28.
- *Incised.* Irregularly and deeply cut, as the edge of a leaf.
- *Incurved.* Gradually curving inward.
- *Indefinite annual growth*, 30.
- *Indehiscent.* Not splitting open.
- *Inflexed.* Bent inward, 33.
- *Involucre.* A whorl or set of bracts around a flower, a cluster of flowers, or fruit, 27.
- *Involute.* Rolled inward from the edges, 34.
- *Irregular.* Said of a flower which has its corolla of different sized, shaped, or colored pieces, 25.

- *Kernel.* The substance contained within the shell of a nut or the stone of a fruit.
- *Key.* A fruit furnished with a wing, or leaf-like expansion, 28.
- *Kidney-shaped.* Broadly heart-shaped, with the apex and basal notch somewhat rounded.

- *Lacerated.* With a margin irregularly notched or apparently torn.
- *Laciniate.* Cut into narrow lobes; slashed.

- *Lance-shaped. Lanceolate.* Like a lance-head in shape, 21.
- *Leaf*, 17; arrangement of leaves, 18; bases of, 21; forms of, 20; kinds of, 19; margins of, 22; parts of, 19; points of, 22; veining, 19.
- *Leaflet.* A separate blade of a compound leaf, 20.
- *Leafstalk.* The stem of a leaf; petiole, 19.
- *Legume.* A pea-like pod, 28.
- *Lensform. Lenticular.* Thickest in the center, with the edges somewhat sharp; like a double-convex lens.
- *Linear.* Long and narrow, with the edges about parallel, 20. [Pg 208]
- *Lobe.* The separate, projecting parts of an irregularly edged leaf if few in number, 22.
- *Lobed.* Having lobes along the margin, 22.

- *Margin of leaves*, 22.
- *Medullary rays*, 13.
- *Membranous.* Thin and rather soft, and more or less translucent, 23.
- *Midrib.* The central or main rib of a leaf, 19.
- *Monœcious.* With both pistillate and staminate flowers on the same plant, 25.
- *Monopetalous.* With the corolla more or less grown together at the base; gamopetalous, 25.
- *Mucronate.* Tipped with a short abrupt point, 22.
- *Multiple roots*, 9.

- *Nerved.* Parallel-veined, as the leaves of some trees, 20.
- *Netted-veined.* With branching veins, forming a network as in the leaves of most of our trees, 20.
- *Node.* The part of a stem to which a leaf is attached, 18.
- *Nut.* A hard, unsplitting, usually one-seeded fruit, 27.
- *Nutlet.* A small nut.

- *Obcordate.* Heart-shaped, with the stem at the pointed end, 21, 22.

- *Oblanceolate.* Lanceolate, with the stem at the more pointed end, 21.
- *Oblong.* Two to four times as long as wide, with the sides somewhat parallel, 20.
- *Oblique.* Applied to leaves when the sides are unequal, 21.
- *Obovate.* A reversed ovate, 21.
- *Obovoid.* A reversed ovoid; an egg form, with stem at the smaller end.
- *Obscurely.* Not distinctly; usually needing a magnifying-glass to determine.
- *Obtuse.* Blunt or rounded at tip, 22.
- *Obvolute*, 34.
- *Odd-pinnate.* Pinnate, with an end leaflet, 20.
- *Once-pinnate.* A compound leaf, with but a single series of leaflets along the central stem, 19.
- *Opposite.* With two leaves on opposite sides of a stem at a node, 18.
- *Orbicular.* Circular in outline, 20.
- *Oval.* Broadly elliptical, 20.
- *Ovary.* The part of the pistil of a flower containing the ovules or future seeds.
- *Ovate.* Shaped like a section of an egg, with the broader end near the stem, 21. [Pg 209]
- *Overlapping.* One piece spreading over another.
- *Ovoid.* Ovate or oval in a solid form, like an egg.
- *Ovules.* The parts within the ovary which may form seeds, 25.

- *Palmate.* A compound leaf, with the leaflets all starting from the end of the petiole, 19.
- *Palmately lobed*, 22.
- *Palmately veined.* With three or more main ribs, or veins of a leaf, starting from the base, 20.
- *Panicle.* An open, much branched cluster of flowers or fruit, 26.
- *Pappus.* The down, hairs, or teeth on the end of the fruit in Compositæ, as the thistle-down.

- *Parallel-veined.* With the veins of the leaf parallel; nerved, 20.
- *Parted.* Edge of a blade separated three fourths of the distance to the base or midrib, 23.
- *Pedicel.* The stem of each flower of a cluster, 26.
- *Peduncle.* The stem of a solitary flower, or the main stem of a cluster, 26.
- *Pellucid.* Almost or quite transparent.
- *Peltate.* Applied to a leaf or other part when the stem or stalk is attached within the margin on the side.
- *Pendent.* Hanging downward, 28.
- *Pendulous.* Hanging or drooping.
- *Perfect.* Said of a flower with both stamen and pistil, 25.
- *Petal.* A leaf of the corolla of a flower, 25.
- *Petiole.* The stalk or stem of a leaf, 19.
- *Petiolate.* Said of a leaf which has a stem, 20.
- *Pinnæ.* The first divisions of a bipinnate or tripinnate leaf.
- *Pinnate leaf.* A compound leaf with the leaflets arranged along the sides of the stem, 19.
- *Pinnately lobed*, 22; *Pinnate-veined*, 20.
- *Pinnatifid.* A leaf deeply notched along the sides in a pinnate manner, 23.
- *Pistil.* The central essential organ of a flower, 25.
- *Pistillate.* A flower with pistil but no stamens, 25.
- *Pith*, 12.
- *Plicate.* Folded like a fan, 34.
- *Pod.* A dry dehiscent fruit like that of the pea, 28.
- *Points of leaves*, 22.
- *Pollarding trees*, 31.
- *Pollen.* The dust or fertilizing material contained in the anther, 24.
- *Polypetalous.* Having a corolla of separate petals, 25.
- *Pome.* An apple-like fruit with the seeds in horny cells, 27.
- *Preparation of a collection*, 35. [Pg 210]
- *Pressing plants*, 36.

- *Prickles.* Sharp, spine-like elevations on the bark, leaf or fruit, 28.
- *Primary root*, 10.
- *Pubescent.* Hairy or downy, especially with fine soft hairs or pubescence, 23.
- *Pulp.* The soft flesh of such fruits as the apple or cherry.
- *Punctate.* With translucent glands, 23.
- *Pyramidal.* With sloping sides like a pyramid, but with a circular base; broad-conical.

- *Raceme.* A flower-cluster with one-flowered stems arranged along the peduncle, 26.
- *Radial section of wood*, 35.
- *Radiating ribs.* The ribs of a leaf when several start together at or near the base. A leaf having such ribs is said to be radiately or palmately veined, 20.
- *Rapier-shaped.* Narrow, pointed, and curved like a sword.
- *Recurved* or *reflexed.* Bent backward, 28.
- *Regular.* Said of a flower which has its enveloping organs alike on all sides, 25.
- *Repand.* Wavy-margined, 22.
- *Retuse.* With a slightly notched tip, 22.
- *Revolute.* Rolled backward, as the edges of many leaves, 22, 34.
- *Ribbed.* With prominent ribs, often somewhat parallel.
- *Ribs.* The strong veins of a leaf, 19.
- *Root*, 9.
- *Rugous.* Having an irregularly ridged surface, 23.

- *Samara.* A winged fruit; a key fruit, 28.
- *Sap-wood*, 13.
- *Scabrous.* Rough or harsh to the touch, 23.
- *Scale-shaped*, 21.
- *Scarious.* Thin, dry, and membranous, 23.
- *Scattered leaves*, 18.
- *Secondary roots*, 10.

- *Section of wood*, 35.
- *Seedling.* A young plant raised from a seed.
- *Seeds*, 25.
- *Sepal.* A division of a calyx, 25.
- *Serrate.* Having a notched edge, with the teeth pointing forward, 22.
- *Serration.* A tooth of a serrated edge.
- *Serrulate.* Finely serrate, 22.
- *Sessile.* Without stem; sessile leaf, 20; sessile flower, 26. [Pg 211]
- *Sheath.* A tubular envelope.
- *Shoot.* A branch.
- *Shrub.* A bush-like plant; one branching from near the base, 11.
- *Silver grain. Medullary rays*, 13, 36.
- *Simple leaf.* One with but a single blade, 19.
- *Sinuate.* With a margin strongly wavy, 22.
- *Sinuation.* One of the waves of a sinuate edge.
- *Spatulate.* Gradually narrowed downward from a rounded tip.
- *Spike.* An elongated cluster of flowers with the separate blossoms about sessile.
- *Spine.* A sharp, rigid outgrowth from the wood of a stem; sometimes applied to sharp points not so deeply seated which should be considered as prickles, 28.
- *Spinescent* or *spiny.* Having spines, 22, 23.
- *Spray.* A collection of small shoots or branches of a plant.
- *Stamen.* One of the pollen-bearing or fertilizing parts of a flower, 24.
- *Staminate.* Said of flowers which have stamens but no pistil, 25.
- *Stellate.* Branching, star-like.
- *Stems and branches*, 11.
- *Stipules.* Small blades at the base of a leafstalk, 19.
- *Straight-veined.* Feather-veined with the veins straight and parallel, 20.
- *Striate.* Marked with fine longitudinal lines or ridges.

- *Sub.* A prefix applied to many botanical terms, and indicating nearly.
- *Subulate.* Awl-shaped, 21.
- *Succulent.* Thick and fleshy, 23.
- *Suckers.* Shoots from a subterranean part of a plant.
- *Surface of leaves and fruit*, 23.

- *Tangential section of wood*, 35.
- *Tapering.* Gradually pointed; gradually narrowed, 21.
- *Tap-root.* A simple root with a stout tapering body, 9.
- *Terete.* Cylindric, but tapering as the twigs of a tree.
- *Terminal.* Belonging to the extremity of a branch, as a *terminal bud*, 14; or *terminal flower-cluster*, 26.
- *Texture of leaves*, 23.
- *Thyrsus.* A compact, much-branched flower- or fruit-cluster, 26.
- *Tomentose.* Covered with matted, woolly hairs, 23.
- *Toothed.* With teeth or short projections.
- *Tree.* A plant with a woody trunk which does not branch near the ground, 11.
- *Truncate.* With a square end as though cut off, 22. [Pg 212]
- *Twice-pinnate.* Applied to a leaf which is twice divided in a pinnate manner, 20.
- *Twice-serrate*, 22. *Twice-crenate*, 22.
- *Two-ranked.* Applied to leaves when they are flattened out in two ranks on opposite sides of a stem, 18; also applied to spray when it branches out in one plane, 19.

- *Umbel.* A cluster of flowers or fruit having stems of about equal length, and starting from the same point, 26.
- *Umbellate.* Like an umbel.

- *Valvate.* Touching edge to edge, 28.
- *Veining of leaves*, 19.
- *Veinlets.* The most minute framework of a leaf, 19.
- *Veins.* The smaller lines of the framework of a leaf, 19.

- *Wedge-shaped.* Shaped like a wedge; cuneate, 21.
- *Whorl.* In a circle around the stem, as the leaves of a plant, 18.
- *Wings.* A blade or leaf-like expansion bordering a part, as a fruit or stem, 28.
- *Winged.* With wing-like membranes.
- *Winter study of trees*, 29.
- *Wood*, 12. [Pg 213]

INDEX TO PART III.

- Abele-tree, 168.
- Abies, 183-187.
- Acanthopanax, 110.
- Acer, 84-88.
- Acuminate-leaved Clethra, 117.
- Æsculus, 81-83.
- African Cedar, 190.
- Ailanthus, 76.
- Albizzia, 96.
- Alcock's Spruce, 181.
- Alder, 147, 148.
- Alleghany Plum, 98.
- Alnus, 147, 148.
- Alternate-leaved Cornel, 112.
- Amelanchier, 107.
- Anacardiaceæ, 89.
- Angelica-tree, 109.
- Angiospermæ, 62.
- Anonaceæ, 68.
- Apple, 101.
- Aralia, 109, 110.
- Araliaceæ, 109.
- Araucaria, 190.
- Arbor-vitæ, American, 194.
    - Chinese, 194.
    - Eastern, 194.
    - Giant, 194.
    - Hatchet-leaved, 194.
    - Japanese, 196.
    - Weeping, 195.
- Arrow-wood, 114.
- Ash, Black, 124.
    - Blue, 124.
    - European, 124.

- o Flowering, 125.
  - o Green, 123.
  - o Red, 123.
  - o Water, 124.
  - o Weeping, 125.
  - o White, 123.
- Ash-colored Willow, 167.
- Ash-leaved Maple, 89.
- Asimina, 68.
- Aspen, 168.
- Austrian Pine, 175.

- Baccharis, 115.
- Bald Cypress, 192.
- Balm of Gilead, 170.
- Balsam-fir, 183, 184.
- Balsam-poplar, 170.
- Barren Oak, 158.
- Bartram's Oak, 152.
- Basket-oak, 154.
- Basswood, 72, 73.
- Bay, Red, 130.
- Bay Willow, 164, 165.
- Beaked Hazelnut, 149.
- Beaked Willow, 166.
- Bean-trefoil Tree, 92.
- Bear Scrub Oak, 157.
- Beech, American, 161.
  - o Blue, 151.
  - o Cut-leaved, 161.
  - o European, 161.
  - o Purple, 161.
  - o Silver Variegated, 161.
  - o Water, 151.
- Benjamin-bush, 131.
- Betula, 144-147.
- Bhotan Pine, 172.

- Bignoniaceæ, 127.
- Bignonia Family, 127.
- Big Shellbark, 142. [Pg 214]
- Big Tree of California, 192.
- Bilsted, 108.
- Biota, 193.
- Birch, American White, 145.
    - Black, 146.
    - Canoe, 145.
    - Cherry, 146.
    - Cut-leaved, 146.
    - European White, 146.
    - Gray, 145, 146.
    - Hairy-leaved, 146.
    - Paper, 145.
    - Purple-leaved, 146.
    - Pyramidal, 146.
    - Red, 147.
    - River, 147.
    - Sweet, 146.
    - Weeping, 146.
    - Yellow, 146.
- Bird-cherry, 99, 100.
- Bitternut, 143.
- Bixineæ, 67.
- Black Ash, 124.
    - Birch, 146.
    - Cherry, 99.
    - Gum, 112.
    - Haw, 114.
    - Hawthorn, 106.
    - Oak, 156, 158.
    - Pine, 175.
    - Poplar, 170.
    - Scrub Oak, 157.
    - Spruce, 179.
    - Sugar-maple, 86.
    - Walnut, 141.

- o   Willow, 163.
- Blackthorn, 98.
- Blue Ash, 124.
  - o   Beech, 151.
- Bog Willow, 166.
- Bow-wood, 137.
- Box Elder, 89.
  - o   White Oak, 153.
- Boxwood, 133.
- Bristly Locust, 94.
- Brittle Willow, 163.
- Broom-hickory, 143.
- Buckeye, 82, 83.
- Buckthorn, California, 80.
  - o   Carolina, 79.
  - o   Common, 79.
  - o   Southern, 119.
  - o   Woolly-leaved, 118.
- Buckthorn Family, 79.
- Buffalo-berry, 132.
- Bullace Plum, 98.
- Bumelia, 118, 119.
- Burning-bush, 78.
- Bur-Oak, 153.
- Butternut, 140.
- Buttonwood, 139.
- Buxus, 132, 133.

- Calico-bush, 116.
- California Buckthorn, 80.
  - o   Maple, 86.
- Camellia Family, 69.
- Canoe Birch, 145.
- Caprifoliaceæ, 113.
- Caragana, 92.
- Carolina Buckthorn, 79.

- o Poplar, 169.
- Carpinus, 150, 151.
- Carya, 141-144.
- Cashew Family, 89.
- Castanea, 159, 160.
- Catalpa, 128, 129.
- Caucasian Planer-tree, 136.
- Cedar, African, 190.
  - o Deodar, 190.
  - o Indian, 190.
  - o Japan, 198.
  - o Lebanon, 189.
  - o Mt. Atlas, 190.
  - o Red, 199.
  - o Silver, 190.
  - o White, 194, 195.
- Cedrela, 76. [Pg 215]
- Cedrus, 189, 190.
- Celastraceæ, 78.
- Celtis, 136, 137.
- Cembra Pine, 173.
- Cephalonian Silver Fir, 185.
- Cephalotaxus, 200.
- Cercidiphyllum, 67.
- Cercis, 94.
- Chaste-tree, 130.
- Cherry, 99, 100.
- Cherry Birch, 146.
- Cherry, Cornelian, 111.
- Chestnut, 160.
- Chestnut-oak, 154, 155.
- Chickasaw Plum, 98.
- Chile Pine, 190.
- China-tree, 75.
- Chinese Arbor-vitæ, 194.
  - o Cedrela, 76.
  - o Cork-tree, 74.

- o Honey-locust, 96.
- o Parasol, 72.
- o Sumac, 91.
- o White Magnolia, 65.
- Chinquapin, 160.
- Chionanthus, 126.
- Choke-cherry, 100.
- Cilician Silver Fir, 186.
- Cladrastis, 93.
- Clammy Locust, 94.
- Clerodendron, 129.
- Clethra, 117, 118.
- Club, Hercules', 109.
- Cockspur Thorn, 104.
- Coffee-tree, Kentucky, 95.
- Colchicum-leaved Maple, 87.
- Compositæ, 115.
- Coniferæ, 170.
- Cork-bark Maple, 87.
- Cork Elm, 134.
- Cork-tree, Chinese, 74.
- Cornaceæ, 110.
- Cornel, 111, 112.
- Cornelian Cherry, 111.
- Cornus, 110-112.
- Corsican Pine, 175.
- Corylus, 149.
- Cottonwood, 169.
- Cow-oak, 154.
- Crab-apple, 101.
- Crack-willow, 163.
- Cranberry-tree, 114.
- Crape-myrtle, 109.
- Cratægus, 103-106.
- Crisped-leaved Elm, 134.
- Cryptomeria, 198.

- Cucumber-tree, 63, 64.
- Cunninghamia, 191.
- Cupuliferæ, 144.
- Custard-apple Family, 68.
- Cut-leaved Birch, 146.
    - Alder, 148.
- Cypress, Bald, 192.
    - Lawson's, 196.
    - Nootka Sound, 195.
    - Southern, 192.

- Dahoon Holly, 77.
- Date-plum, 120.
- Deodar Cedar, 190.
- Devil-wood, 125.
- Diospyros, 119, 120.
- Dogwood, Flowering, 111.
    - Poison, 90.
- Dotted-fruited Hawthorn, 106.
- Double Spruce, 179.
- Downy-leaved Poplar, 169.
- Dwarf Chestnut-oak, 155.
- Dwarf Mountain Sumac, 90.

- Ear-leaved Umbrella-tree, 64.
- Eastern Spruce, 181.
- Ebenaceæ, 119.
- Ebony Family, 119.
- Elæagnaceæ, 131.
- Elæagnus, 131, 132.
- Elder-leaved Mountain Ash, 102.
- Elder, Poison, 90. [Pg 216]
- Elm, American, 135.
    - Cork, 134.
    - Crisped-leaved, 134.
    - English, 134.

- o Field, 134.
  - o Kiaka, 136.
  - o Red, 134.
  - o Rock, 134.
  - o Scotch, 134.
  - o Slippery, 134.
  - o Wahoo, 135.
  - o Weeping, 134.
  - o White, 135.
  - o White-margined, 134.
  - o Winged, 135.
  - o Witch, 134.
- English Elm, 134.
  - o Cherry, 99.
  - o Hawthorn, 104.
  - o Maple, 87.
  - o Oak, 158.
  - o Walnut, 141.
- Ericaceæ, 116.
- Euonymus, 78.
- Euphorbiaceæ, 132.

- Fagus, 160, 161.
- Fate-tree, 129.
- Field Elm, 134.
- Figwort Family, 127.
- Filbert, 149.
- Fir, Balsam, 183, 184.
  - o Cephalonian Silver, 185.
  - o Cilician Silver, 186.
  - o European Silver, 187.
  - o Fraser's Balsam, 184.
  - o Great Silver, 185.
  - o Japan Silver, 184.
  - o Noble Silver, 187.
  - o Nordmann's Silver, 184.
  - o Pinsapo, 186.
  - o Scotch, 177.

- o	Siberian Silver, 185.
  - o	Silver, 184-187.
  - o	Southern Balsam, 184.
  - o	White, 186.
- Flowering Ash, 125.
  - o	Dogwood, 111.
- Four-winged Silverbell Tree, 121.
- Fraser's Balsam-fir, 184.
- Fraxinus, 122-125.
- French Tamarisk, 69.
- Fringe-tree, 126.

- Garden Plum, 99.
  - o	Red Cherry, 99.
- Garland Crab-apple, 101.
- Giant Arbor-vitæ, 194.
  - o	Tree Lilac, 126.
- Ginkgo-tree, 201.
- Gleditschia, 95, 96.
- Goat-willow, 166.
- Golden-chain, 92.
- Golden Larch, 189.
- Gordonia, 70.
- Gray Birch, 145, 146.
  - o	Pine, 178.
  - o	Willow, 167.
- Great Laurel, 117.
- Great-leaved Magnolia, 64.
- Great Silver Fir, 185.
  - o	Tree of California, 192.
- Green Ash, 123.
- Groundsel-tree, 115.
- Gum, Black, 112.
  - o	Sour, 112, 113.
  - o	Sweet, 108.
- Gymnocladus, 95.
- Gymnospermæ, 170.

- Hackberry, 136.
- Hackmatack, 188.
- Halesia, 121.
- Hamamelideæ, 107.
- Hamamelis, 107.
- Hatchet-leaved Arbor-vitæ, 194.
- Haw, Black, 114. [Pg 217]
  - Summer, 106.
  - Yellow, 106.
- Hawthorn, Black, 106.
  - Dotted-fruited, 106.
  - English, 104.
  - Pear, 106.
  - Tall, 105.
- Hazel, 149.
- Hazelnut, 149.
- Heart-leaved Alder, 148.
  - Willow, 165.
- Heath Family, 116.
- Heavy-wooded Pine, 174.
- Hemlock, Common, 182.
  - Ground, 199.
  - Japan, 182.
  - Mountain, 182.
- Hercules'-Club, 109.
- Hibiscus, 71.
- Hickory, Big Shellbark, 142.
  - Broom, 143.
  - Shagbark, 142.
  - Shellbark, 142.
  - Swamp, 143.
  - White-heart, 142.
- Himalayan Spruce, 181.
- Hoary Alder, 147.
- Holly, 77.
- Holly Family, 77.
- Honey-locust, 95, 96.

- Honeysuckle Family, 113.
- Hop-hornbeam, 150.
- Hop-tree, 74.
- Hornbeam, 151.
- Horse-chestnut, 81, 82.
- Horse-sugar, 122.
- Hovenia, 80.

- Idesia, 67.
- Ilex, 77, 78.
- Ilicineæ, 77.
- Imperial Paulownia, 127.
- Indian Bean, 128.
    - Cedar, 190.
- Irish Juniper, 199.
- Iron-wood, 150.

- Japan Arbor-vitæ, 196.
    - Cedar, 198.
    - Hemlock, 182.
    - Larch, 188.
    - Lilac, 126.
    - Magnolia, 65.
    - Maple, 88.
    - Persimmon, 120.
    - Planer-tree, 136.
    - Pine, 176.
    - Podocarpus, 201.
    - Silver Fir, 184.
    - Storax, 120.
- Jersey Pine, 177.
- Judas-tree, 94.
- Juglandaccæ, 140.
- Juglans, 140, 141.
- Jujube, 80.
- Juniper, 198, 199.
- Juniperus, 198, 199.

- Kalmia, 116.
- Katsura-tree, 67.
- Kentucky Coffee-tree, 95.
- Kiaka Elm, 136.
- Kilmarnock Willow, 166.
- Kingnut, 142.
- Kœlreuteria, 83.

- Laburnum, 92.
- Lagerstrœmia, 109.
- Lambert's Pine, 172.
- Larch, American, 188.
  - European, 188.
  - Golden, 189.
  - Japan, 188.
- Large-flowered Magnolia, 63.
- Large-leaved Maple, 86.
- Large-toothed Aspen, 168.
- Large Tupelo, 113.
- Larix, 187-189. [Pg 218]
- Lauraceæ, 130.
- Laurel, 116, 117.
- Laurel Family, 130.
- Laurel-leaved Willow, 165.
- Laurel-oak, 158.
- Lawson's Cypress, 196.
- Lebanon Cedar, 189.
- Leguminosæ, 92.
- Lilac, 126.
- Linden, 72, 73.
- Linden Family, 72.
- Lindera, 131.
- Liquidambar, 108.
- Liriodendron, 66.
- Live-oak, 155.
- Loblolly Bay, 70.

- o Pine, 174.
- Locust, Bristly, 94.
  - o Clammy, 94.
  - o Common, 93.
  - o Honey, 95, 96.
- Lombardy Poplar, 169.
- Long-leaved Pine, 174.
  - o Willow, 167.
- Long-racemed Buckeye, 83.
- Lonicera, 115.
- Loosestrife Family, 108.
- Lythraceæ, 108.

- Maclura, 137.
- Madeira Nut, 141.
- Magnolia, Chinese White, 65.
  - o Great-leaved, 64.
  - o Japan, 65.
  - o Large-flowered, 63.
  - o Purple Japan, 66.
  - o Southern Evergreen, 63.
  - o Swamp, 63.
  - o Thurber's Japan, 66.
- Magnoliaceæ, 62.
- Magnolia Family, 62.
- Mallow Family, 71.
- Malvaceæ, 71.
- Maple, Ash-leaved, 89.
  - o California, 86.
  - o Colchicum-leaved, 87.
  - o Cork-bark, 87.
  - o English, 87.
  - o Japan, 88.
  - o Large-leaved, 86.
  - o Mountain, 84.
  - o Norway, 87.
  - o Palmate-leaved, 88.

o   Red, 85.
o   Rock, 86.
o   Round-leaved, 88.
o   Silver, 85.
o   Striped, 85.
o   Sugar, 86.
o   Sycamore, 86.
o   Tartarian, 88.
o   Vine, 88.
o   White, 85.
- Masson's Pine, 175.
- Melia, 75.
- Meliaceæ, 75.
- Melia Family, 75.
- Mockernut, 142, 143.
- Morello Cherry, 99.
- Morus, 137, 138.
- Mossy-cup Oak, 153.
- Mountain Ash, 102, 103.
  - o   Hemlock, 182.
  - o   Laurel, 116.
  - o   Maple, 84.
  - o   Pine, 173, 177.
  - o   Sumac, 90.
- Mount Atlas Cedar, 190.
- Mulberry, 138.
  - o   Paper, 138.
- Myrtle, Crape, 109.

- Narrow-leaved Crab-apple, 101.
- Necklace-poplar, 169.
- Negundo, 88, 89.
- Noble Silver Fir, 187.
- Nootka Sound Cypress, 195.
- Nordmann's Silver Fir, 184. [Pg 219]
- Northern Prickly Ash, 73.
  - o   Scrub Pine, 178.

- Norway Maple, 87.
  - Spruce, 180.
- Nut, Bitter, 143.
  - Hickory, 142, 143.
  - King, 142.
  - Mocker, 142, 143.
  - Pecan, 144.
  - Pig, 143.
- Nut-pine, 178.
- Nyssa, 112, 113.

- Oak, American White, 153.
  - Barren, 158.
  - Bartram's, 152.
  - Basket, 154.
  - Bear Scrub, 157.
  - Black, 156, 158.
  - Black Scrub, 157.
  - Box White, 153.
  - Bur, 153.
  - Chestnut, 154, 155.
  - Cow, 154.
  - English, 158.
  - Laurel, 158.
  - Live, 155.
  - Mossy-cup, 153.
  - Pin, 156.
  - Post, 153, 154.
  - Pyramidal, 159.
  - Quercitron, 156.
  - Red, 156.
  - Rough, 153.
  - Scarlet, 156.
  - Scrub, 157.
  - Shingle, 158.
  - Spanish, 156, 157.
  - Swamp, 154, 156.
  - Turkey, 159.

- o Water, 157.
- o Weeping, 159.
- o White, 153, 154.
- o Willow, 158.
- o Yellow, 155, 156.
- Oak Family, 144.
- Oak-leaved Alder, 148.
  - o Mountain-ash, 102.
- Ohio Buckeye, 82.
- Old-field Pine, 174.
- Oleaceæ, 122.
- Oleaster Family, 131.
- Olive Family, 122.
- Orange, Osage, 137.
- Oriental Plane, 139.
  - o Spruce, 181.
- Osage Orange, 137.
- Osmanthus, 125.
- Ostrya, 150.
- Oxydendrum, 116.

- Palmate-leaved Japan Maple, 88.
- Papaw, 68.
- Paper Birch, 145.
  - o Mulberry, 138.
- Parsley-leaved Thorn, 105.
- Paulownia, 127.
- Peach, 97.
- Pear Hawthorn, 106.
- Pear-tree, 101.
- Pea-tree, 92.
- Pecan-nut, 144.
- Pepperbush, 117, 118.
- Pepperidge, 112.
- Persea, 130.
- Persimmon, 119, 120.

- Phellodendron, 74.
- Picea, 179-181.
- Pignut, 143.
- Pine, Austrian, 175.
  - Bhotan, 172.
  - Black, 175.
  - Cembra, 173.
  - Chile, 190.
  - Corsican, 175.
  - Gray, 178.
  - Heavy-wooded, 174.
  - Japan, 176. [Pg 220]
  - Jersey, 177.
  - Lambert's, 172.
  - Loblolly, 174.
  - Long-leaved, 174.
  - Masson's, 175.
  - Mountain, 173, 177.
  - Nut, 178.
  - Old-field, 174.
  - Piñon, 178.
  - Pitch, 174.
  - Red, 176.
  - Scotch, 177.
  - Scrub, 177, 178.
  - Stone, 173.
  - Sugar, 172.
  - Swiss Stone, 173.
  - Table-Mountain, 177.
  - Twisted-branched, 177.
  - Umbrella, 191.
  - Weymouth, 172.
  - White, 172, 173.
  - Yellow, 174,176.
- Pine Family, 170.
- Pin-oak, 156.
- Piñon Pine, 178.
- Pinsapo Fir, 186.

- Pitch-pine, 174.
- Pinus Austriaca, 175.
    - Banksiana, 178.
    - Cembra, 173.
    - contorta, 177.
    - densiflora, 176.
    - edulis, 178.
    - excelsa, 172.
    - flexilis, 173.
    - inops, 177.
    - Lambertiana, 172.
    - Laricio, 175.
    - Massoniana, 175.
    - mitis, 176.
    - monophylla, 178.
    - monticola, 173.
    - palustris, 174.
    - ponderosa, 174.
    - pungens, 177.
    - resinosa, 176.
    - rigida, 174.
    - strobus, 172.
    - sylvestris, 177.
    - Tæda, 174.
- Plane, Oriental, 139.
- Planera, 135, 136.
- Planer-tree, 136.
- Plane-tree Family, 139.
- Platanaceæ, 139.
- Platanus, 139.
- Plum, 98, 99.
- Plum, Date, 120.
- Podocarpus, 200, 201.
- Poison Dogwood, 90.
    - Elder, 90.
    - Sumac, 90.
- Pomegranate-tree, 108.
- Populus, 167-170.

- Poplar, Balsam, 170.
    - Black, 170.
    - Carolina, 169.
    - Downy-leaved, 169.
    - Lombardy, 169.
    - Necklace, 169.
    - White, 168.
- Post-oak, 153, 154.
- Prickly Ash, 73, 74.
- Pride of India, 75.
- Prunus, 97-100.
- Ptelea, 74.
- Pterostyrax, 121.
- Pulse Family, 92.
- Punica, 108.
- Purple Japan Magnolia, 66.
- Purple-leaved Birch, 146.
- Purple Willow, 165.
- Pyramidal Birch, 146.
    - Oak, 159.
- Pyrus, 100-103.

- Quaking-asp, 168.
- Quassia Family, 76. [Pg 221]
- Quercitron Oak, 156.
- Quercus alba, 153.
    - aquatica, 157.
    - bicolor, 154.
    - Cerris, 159.
    - coccinea, 156.
    - falcata, 157.
    - fastigiata, 159.
    - heterophylla, 152.
    - ilicifolia, 157.
    - imbricaria, 158.
    - lyrata, 154.
    - macrocarpa, 153.

- o Michauxii, 154.
  - o Muhlenbergii, 155.
  - o nigra, 158.
  - o palustris, 156.
  - o pedunculata, 159.
  - o pendula, 159.
  - o Phellos, 152, 158.
  - o prinoides, 155.
  - o Prinus, 154.
  - o Robur, 158.
  - o rubra, 152, 156.
  - o sessiliflora, 159.
  - o stellata, 153.
  - o tinctoria, 156.
  - o virens, 155.
- Quince-tree, 102.

- Rabbit-berry, 132.
- Red Ash, 123.
  - o Bay, 130.
  - o Birch, 147.
  - o Buckeye, 82.
  - o Cedar, 199.
  - o Cherry, 99.
  - o Elm, 134.
  - o Horse-chestnut, 82.
  - o Maple, 85.
  - o Mulberry, 138.
  - o Oak, 156.
  - o Pine, 176.
  - o Plum, 98.
- Redbud, 94.
- Red-leaved Alder, 148.
- Redwood, 193.
- Retinospora, 193, 196, 197.
- Rhamnaceæ, 79.
- Rhamnus, 79, 80.
- Rhododendron, 117.

- Rhus, 89-91.
- River Birch, 147.
- Robinia, 93, 94.
- Rock Elm, 134.
  - Maple, 86.
- Rosaceæ, 97.
- Rose-acacia, 94.
- Rose Family, 97.
- Rough Oak, 153.
- Round-leaved Maple, 88.
- Rowan-tree, 103.
- Rue Family, 73.
- Rutaceæ, 73.

- Salicaceæ, 161.
- Salisburia, 201.
- Salix Alba, 164.
  - amygdaloides, 163.
  - angustata, 165.
  - annularis, 164.
  - Babylonica, 164.
  - caprea, 166.
  - cinerea, 167.
  - cordata, 165.
  - decipiens, 164.
  - discolor, 166.
  - falcata, 163.
  - fragilis, 163.
  - longifolia, 167.
  - lucida, 164.
  - myricoides, 165.
  - nigra, 163.
  - pentandra, 165.
  - purpurea, 165.
  - rigida, 165.
  - rostrata, 166.
  - rufescens, 165. [Pg 222]

- o   Russelliana, 164
- o   viridis, 164.
- o   vitellina, 164.
- Sapindaceæ, 81.
- Sapodilla Family, 118.
- Sapotaceæ, 118.
- Sassafras, 130, 131.
- Scarlet-fruited Thorn, 104.
- Scarlet Oak, 156.
- Sciadopitys, 191.
- Scotch Elm, 134.
    - o   Fir, 177.
    - o   Pine, 177.
- Scrophulariaceæ, 127.
- Scrub Oak, 157.
    - o   Pine, 177, 178.
- Seaside Alder, 148.
- Sequoia, 192, 193.
- Service-berry, 107.
- Shad-bush, 107.
- Shagbark Hickory, 142.
- Sheep-berry, 114.
- Shellbark Hickory, 142.
- Shepherdia, 132.
- Shingle Oak, 158.
- Shining Willow, 164.
- Shrubby Trefoil, 74.
- Siberian Cornel, 111.
    - o   Silver Fir, 185.
- Silk-tree, 96.
- Silverbell-tree, 121.
- Silver Cedar, 190.
    - o   Fir, 184-187.
    - o   Maple, 85.
    - o   Spruce, 181.
- Silver-leaved Elæagnus, 132.
- Simarubaceæ, 76.

- Single Spruce, 179.
- Slippery Elm, 134.
- Sloe, 98.
- Smoke-tree, 91.
- Smooth Alder, 148.
  - Sumac, 90.
- Soapberry Family, 81.
- Sorrel-tree, 116.
- Sour Gum, 112, 113.
- Sourwood, 116.
- Southern Cypress, 192.
- Spanish Oak, 156, 157.
- Speckled Alder, 147.
- Spice-bush, 131.
- Spindle-tree, 78.
- Spruce, Alcock's, 181.
  - Black, 179.
  - Double, 179.
  - Eastern, 181.
  - Himalayan, 181.
  - Norway, 180.
  - Oriental, 181.
  - Silver, 181.
  - Single, 179.
  - Tiger's-tail, 180.
  - White, 179.
- Spurge Family, 132.
- Stag-horn Sumac, 90.
- Sterculia, 71.
- Sterculiaceæ, 71.
- Stone-pine, 173.
- Storax, 120.
- Storax Family, 120.
- Striped Maple, 85.
- Stuartia, 69, 70.
- Styracaceæ, 120.
- Styrax, 120.

- Sugarberry, 136.
- Sugar Maple, 86.
    - Pine, 172.
- Sumac, 90, 91.
- Summer Haw, 106.
- Swamp Hickory, 143.
    - Magnolia, 63.
    - Oak, 156.
    - Post-oak, 154.
    - White Oak, 154.
- Sweet Bay, 63.
    - Birch, 146.
    - Buckeye, 82.
    - Gum, 108. [Pg 223]
    - Pepper-bush, 117, 118.
    - Viburnum, 114.
- Sweetleaf, 122.
- Swiss Stone-pine, 173.
- Sycamore, American, 139.
- Sycamore-maple, 86.
- Symplocos, 122.
- Syringa, 126.

- Table-Mountain Pine, 177.
- Tacamahac, 170.
- Tamarack, 188.
- Tamariscineæ, 68.
- Tamarisk, 69.
- Tamarix, 69.
- Tartarian Honeysuckle, 115.
    - Maple, 88.
- Taxodium, 192.
- Tea Family, 69.
- Ternstrœmiaceæ, 69.
- Thorn, 104, 105.
- Thurber's Japan Magnolia, 66.
- Thuya, 193, 194.

- Thuyopsis, 193.
- Tiger's-tail Spruce, 180.
- Tilia, 72, 73.
- Tiliaceæ, 72.
- Toothache-tree, 73.
- Torreya, 200.
- Tree Hibiscus, 71.
- Tree of Heaven, 76.
- Trefoil, 74.
- Tsuga, 182.
- Tulip-tree, 66.
- Tupelo, 113.
- Turkey Oak, 159.

- Ulmus, 133-135.
- Umbrella-pine, 191.
- Umbrella-tree, 65.
- Urticaceæ, 133.

- Venetian Sumac, 91.
- Verbenaceæ, 129.
- Viburnum, 113, 114.
- Vine Maple, 88.
- Vitex, 129, 130.

- Wahoo, 78, 135.
- Walnut, 140, 141.
- Walnut Family, 140.
- Washington Thorn, 105.
- Water Ash, 124.
    - Beech, 151.
    - Locust, 96.
    - Oak, 157.
- Weeping Ash, 125.
    - Birch, 146.
    - Elm, 134.

- o Oak, 159.
- o Willow, 164.
- White Ash, 123.
  - o Basswood, 73.
  - o Birch, 145, 146.
  - o Cedar, 194, 195.
  - o Elm, 134, 135.
  - o Fir, 186.
  - o Maple, 85.
  - o Mulberry, 138.
  - o Oak, 153, 154.
  - o Poplar, 168.
  - o Spruce, 179.
  - o Willow, 164.
- White-heart Hickory, 142.
- Whitewood, 72.
- Willow, American Bay, 164.
  - o Ash-colored, 167.
  - o Bay, 164, 165.
  - o Beaked, 166.
  - o Black, 163.
  - o Bog, 166.
  - o Brittle, 163.
  - o Crack, 163.
  - o Glaucous, 166.
  - o Goat, 166.
  - o Gray, 167.
  - o Heart-leaved, 165.
  - o Kilmarnock, 166. [Pg 224]
- Willow, Laurel-leaved, 165.
  - o Long-leaved, 167.
  - o Purple, 165.
  - o Shining, 164.
  - o Weeping, 164.
  - o White, 164.
- Willow Family, 161.
- Willow-oak, 158.
- Winged Elm, 135.

- Witch-elm, 134.
- Witch-hazel, 107.
- Witch-hazel Family, 107.

- Xanthoxylum, 73.

- Yellow-barked Oak, 156.
- Yellow Birch, 146.
  - Cucumber-tree, 64.
  - Haw, 106.
  - Plum, 98.
- Yellow-wood, 93.
- Yew, 199.
- Yulan, 65.

- Zizyphus, 80.